Springer Tracts in Advanced Robotics

Volume 64

Editors: Bruno Siciliano · Oussama Khatib · Frans Groen

Alejandro Dizan Vasquez Govea

Incremental Learning for Motion Prediction of Pedestrians and Vehicles

Springer

Professor Bruno Siciliano, Dipartimento di Informatica e Sistemistica, Università di Napoli Federico II, Via Claudio 21, 80125 Napoli, Italy, E-mail: siciliano@unina.it

Professor Oussama Khatib, Artificial Intelligence Laboratory, Department of Computer Science, Stanford University, Stanford, CA 94305-9010, USA, E-mail: khatib@cs.stanford.edu

Professor Frans Groen, Department of Computer Science, Universiteit van Amsterdam, Kruislaan 403, 1098 SJ Amsterdam, The Netherlands, E-mail: groen@science.uva.nl

Author

Dr. Alejandro Dizan Vasquez Govea
ETH Zürich
Autonomous Systems Lab
Tannenstr. 3
8092 Zurich
Switzerland
dichodaemon@gmail.com

ISBN 978-3-642-13641-2 e-ISBN 978-3-642-13642-9

DOI 10.1007/978-3-642-13642-9

Springer Tracts in Advanced Robotics ISSN 1610-7438

Library of Congress Control Number: 2010928270

Typeset & Cover Design: Scientific Publishing Services Pvt. Ltd., Chennai, India.

Printed on acid-free paper

5 4 3 2 1 0

springer.com

STAR (Springer Tracts in Advanced Robotics) has been promoted under the auspices of EURON (European Robotics Research Network)

To Lota, my son, for being unpredictable

Foreword

Robotics is undergoing a major transformation in scope and dimension. From a largely dominant industrial focus, robotics is rapidly expanding into human environments and vigorously engaged in its new challenges. Interacting with, assisting, serving, and exploring with humans, the emerging robots will increasingly touch people and their lives.

Beyond its impact on physical robots, the body of knowledge robotics has produced is revealing a much wider range of applications reaching across diverse research areas and scientific disciplines, such as: biomechanics, haptics, neurosciences, virtual simulation, animation, surgery, and sensor networks among others. In return, the challenges of the new emerging areas are proving an abundant source of stimulation and insights for the field of robotics. It is indeed at the intersection of disciplines that the most striking advances happen.

The Springer Tracts in Advanced Robotics (STAR) is devoted to bringing to the research community the latest advances in the robotics field on the basis of their significance and quality. Through a wide and timely dissemination of critical research developments in robotics, our objective with this series is to promote more exchanges and collaborations among the researchers in the community and contribute to further advancements in this rapidly growing field.

The monograph written by Alejandro Dizan Vasquez Govea focuses on the practical problem of moving in a cluttered environment with pedestrians and vehicles. A framework based on Hidden Markov models is developed to learn typical motion patterns which can be used to predict motion on the basis of sensor data. All the theoretical results have been implemented and validated with experiments, using both real and simulated data.

Remarkably, the monograph is based on the author's doctoral thesis, which received the prize of the Eight Edition of the EURON Georges Giralt PhD Award devoted to the best PhD thesis in Robotics in Europe. A very fine addition to STAR!

Naples, Italy
January 2010

Bruno Siciliano
STAR Editor

Preface

The main subject of this book is motion prediction. The problem is approached from the hypothesis that the dynamic and kinematic properties of objects such as pedestrians and vehicles do not suffice to predict their motion in the long term. Instead, the work presented here, inscribes itself in a new family of approaches which assume that, in a given environment, objects do not move at random, but engage in typical motion patterns, which may be learned and then applied to predict motion on the basis of sensor data. In this context, this book focuses in three fundamental questions: modeling, learning and prediction.

Modeling

The approach proposed in this book is based on Hidden Markov Models, a probabilistic framework, which is used as a discrete approximation to represent the continuous state-space in which motion takes place. The main originality of the approach lies in modeling explicitly the intentions which are at the origin of typical motion patterns. This is achieved through the use of an extended space, which appends the state that the object intends to reach to the other "classic" state variables, such as position or velocity.

Learning

The main problem of existing approaches lies is that they split up model learning and utilization in two distinct stages: in a first phase, the model is learned from data; then, it is used to predict. This principle is difficult to apply to real situations, because it requires at least one example of every possible typical pattern to be available during the learning phase.

To address this problem, this book proposes a novel extension to Hidden Markov Models which effectively enables "lifelong learning" to happen at the same time that the model is used to predict. This extension incrementally builds a topological map – representing the model's structure – and reestimates the model's parameters. The approach is intended to be general, and – in principle –it could be used in other application domains such as gesture recognition or automatic landmark extraction.

Prediction

In the proposed approach, prediction is carried on by using exact Bayesian inference algorithms, which are able to provide real-time performance thanks to the properties of the structure which has been learned. In particular, the time complexity of inference is reduced from $O(N^2)$ to $O(N)$ with respect to the number of discrete states in the system.

Our approach has been implemented and validated with experiments, using both real and synthetic data. Real data has been obtained on a visual tracking system installed over a parking lot. For synthetic data, a simulator has been developed in order to facilitate the realization of controlled tests and the study of our approach in large environments.

Acknowledgements

This book is the result of four years of research during my PhD program at the Institut National Polytechnique de Grenoble and its realization has been possible thanks to the contribution of many people. First of all, I would like to thank Christian Laugier and Thierry Fraichard, my thesis directors, for their guidance, their support, and for bearing with my idiosyncratic work methods.

I would like to thank the members of my jury; in particular Roland Siegwart and Michel Devy, my reviewers, who took the time and energy to point out my errors and omissions. My thanks also go to my examiners Wolfram Burgard, Hiromichi Yanagihara and Olivier Aycard for the interest they manifested on my work. I want to thank Augustin Lux for honoring me by presiding my jury.

I'd also like to thank all of my colleagues at team eMotion with whom I had the pleasure of working over the years. Special thanks go to Thiago Bellardi and Agostino Martinelli for carefully reading my manuscript and for the many corrections and suggestions they made.

I am very grateful to Hannah Dee from the University of Leeds for kindly sharing her data with me.

I would like to thank Fernando Ramos, whose help and advice are at the very origin of this adventure.

And last, but not least, I want to express all my gratitude to Barbara, my partner, for her support, friendship, and for her enormous patience during all this period, thank you Babalucalas!

Contents

Part II: State of the Art

Part III: Proposed Approach

Part IV: Experiments

List of Figures

List of Algorithms

Chapter 1
Introduction

Life is a series of collisions with the future

JOSE ORTEGA Y GASSET

1.1 Motivation

Motion planning for dynamic environments is a very active research domain in robotics. Owing to the fact that the problem is NP-Hard [Reif and Sharir, 1985], most research efforts have been directed towards coping with planning complexity.

There is, however, another critical aspect of the problem which is often overlooked: motion planning algorithms need to know in advance how the objects that populate the environment will move. In practice, this knowledge is seldom available, making it necessary to resort to prediction: gathering information about moving objects through a sensor device (*eg* radar, visual tracker, laser scanner) and then feeding this information into a mathematical model of the object's motion in order to obtain an estimation of its future state (*eg* position, velocity).

Until recently, most motion prediction techniques have been based on kinematic or dynamic models that describe how the state of an object evolves over time when it is subject to a given control (*eg* acceleration) (cf. [Zhu, 1990]). These approaches proceed by estimating the state, using techniques such as the Kalman Filter [Kalman, 1960], and then applying the estimate to its motion equations in order to get state predictions.

Although these techniques are able to produce very good short-term predictions, their performance degrades quickly as they try to see further away in the future. This is especially true for humans, vehicles, robots, animals and the like, which are able to modify their trajectory according to factors

A.D. Vasquez: Incre. Learn. for Motion Predict. of Pedest. & Vehicles, STAR 64, pp. 1–7.
springerlink.com

(*eg* perception, internal state, intentions, etc.) which are not described by their kinematic or dynamic properties.

To address this issue, a different family of approaches has emerged in the last decade. It is based on the idea that, for a given environment, moving objects tend to follow typical motion patterns that depend on the objects' nature and the structure of the environment. These approaches operate in two stages:

1. *Learning stage:* observe the moving objects in the workspace in order to identify and build representations of typical motion patterns.
2. *Prediction stage:* use the learned typical motion patterns to predict the future motion of a given object.

Thus, learning consists of observing a given environment in order to construct a representation of every possible motion pattern. But, how long should we observe the environment in order to construct such a "library" of motion patterns? Given the enormous number of possible behaviors the humans may exhibit in all but the simplest environments, there is not a simple answer. This raises an important problem of most current learning techniques [Hu et al., 2004a, Bennewitz et al., 2005, Wang et al., 2006]: they use a "learn then predict" approach, meaning that the system goes through a learning stage where it is presented with an example data set from which it builds its pattern models. Then, the models are "frozen" and the system goes into the prediction stage.

The problem with this approach is that it makes the implicit assumption that there is enough information in the example data set to build a representation of every possible motion pattern, which, as we have shown, is a difficult condition to guarantee. This book proposes a solution to this problem: a lifelong learning approach, where learning and prediction take place in a continuous and parallel fashion, making it possible to refine knowledge on the basis of the same observations that are used to predict motion as it happens.

1.2 Problem Description

It will be assumed hereafter that the input of a learning based algorithm consists of a series of discrete observations (*ie* sensor readings) describing the motion of objects in a given environment. Most of the time we will consider observations to consist of position and velocity information of objects with respect to the plan, nevertheless, sometimes they also include other information like size, orientation, confidence, etc.

In addition to this information, it is assumed that observations are arranged in sequences $O_{1:T} = \{O_1, \cdots, O_T\}$ such that every sequence describes the trajectory of a single object since it is observed for the first time until it leaves the environment or stops moving. For example, in the case of a visual tracker, observation sequences will correspond to complete tracks, from creation until deletion.

A complete learning-based approach, includes the following components:

A *motion pattern model* describing how the object's state evolves as time passes considering that it is engaged in a given motion pattern.
A *learning algorithm* specifying how the model's parameters should be estimated from collected data. For a lifelong learning approach approach, this should ideally comprise the ability to create or delete motion pattern models as learning progresses.
A *prediction algorithm* prescribing the use of learned models and observations to predict future motion. More precisely it should answer the following question: given our current knowledge and the whole history of observations for an object up to the present time t, what will be its state at time $t + H$?

Since uncertainty is inherent to prediction, it seems sensible to use a probabilistic framework and to model motion patterns as stochastic processes. Thus, we have based our work on Hidden Markov Models [Rabiner, 1990], a probabilistic approach which is very popular in the literature of "learn then predict" approaches [*eg* Walter et al., 1999, Makris and Ellis, 2002, Bennewitz et al., 2005] due, at least in part, to the existence of efficient inference and learning algorithms. Nevertheless, the application of Hidden Markov Models (HMM) to lifelong learning, poses additional problems which – as we will see in the following section – are not solved by classical algorithms.

1.2.1 Modeling Motion with Hidden Markov Models

A Hidden Markov Model may be viewed as a graph (fig. 1.1), where nodes represent discrete states (*eg* places in the environment) and edges represent transitions. The topology of this graph (*ie* its edges) and its cardinality (*ie* number of states) are often called the *model's structure*. The model assumes that the state of an object is not perceptible in a direct fashion, instead, there is an *observation probability*, attached to every state, describing how likely it is to obtain a given observation, supposing that the object is in that state. Transitions are not deterministic either, but occur according to a *transition probability*, attached to every edge in the graph. Transition and observation probabilities are often represented as multinomial and Gaussian probabilities, respectively. The set of all the multinomial and Gaussian parameters are known together as the *model's parameters*.

HMM based motion prediction approaches in the literature often model motion patterns as *typical trajectories*, which, in HMMs are represented as chains (*ie* order-two graphs) with directed edges going only in one direction

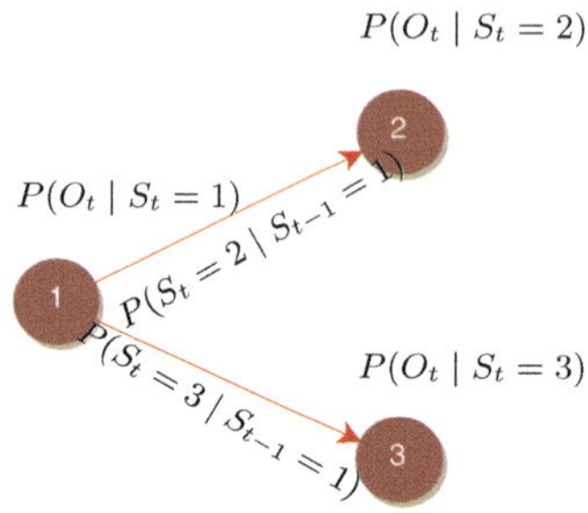

Fig. 1.1 A basic three-state HMM, with transition probabilities attached to edges and observation probabilities attached to nodes.

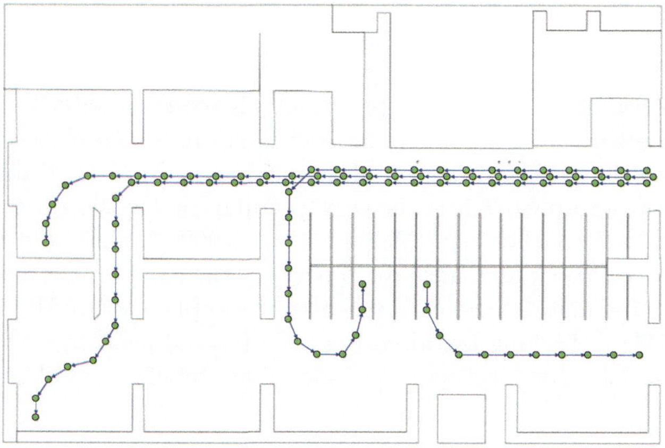

Fig. 1.2 An HMM structure for a parking environment, where motion patterns are represented as non connected chains (order 2 subgraphs) (only a few motion patterns are displayed).

(fig. 1.2). Every such graph constitutes an individual HMM [*eg* Walter et al., 1999]; or, alternatively, they may be represented as a single HMM conformed by many non-connected chains [*eg* Bennewitz et al., 2005].

1.2.2 *Challenges*

When using HMMs as motion models there are actually two learning tasks to be accomplished: a) learn the graph structure, and b) estimate the parameters of the different probability distributions. In most cases, they are roughly equivalent to *identifying* and *representing* motion patterns, respectively.

The standard learning technique for HMMs is the Baum-Welch algorithm Baum et al. [1970] – a specialization of the well known Expectation Maximization algorithm [Dempster et al., 1977] – unfortunately, it is a parameter-only learning algorithm which assumes that the model's structure is known *a priori*. Moreover, in its classical form, the algorithm is not applicable to a lifelong learning approach, because it is designed for off-line use, processing data in batch mode.

Although structure-learning and incremental extensions of the Baum-Welch algorithm exist [Friedman, 1997, Singer and Warmuth, 1996], an algorithm that is able to perform both of them is still needed in order to enable an HMM based lifelong learning approach. Moreover, the algorithm should be able to work in *real time* which, in this context, means delivering predictions in the time elapsed between two sensor readings.

Prediction in HMMs consists of applying Bayesian inference to update a probabilistic estimate (*ie* the belief state) of the object's state with newly gathered observations, and then to project it into the future. This may be done very efficiently on an observation by observation basis, however, depending on the HMM's structure, inference may not be fast enough to enable real-time usage, and it is necessary to apply approximate methods [*eg* Walter et al., 1999].

Another problem comes from the fact that, in real life, persons often change their mind while moving, leading to trajectories which are highly atypical and, at the same time, are composed by well known sub-behaviors. The problem with most current approaches is that they are not able to deal with this "mixed patterns" because typical trajectories are supposed to be completely independent. This leads to bad generalization, since a new motion pattern should be learned for every such a pattern, which is problematic because they will not represent actual behaviors, but an atypical motion which is unlikely to be observed again.

1.3 Contributions

The main contribution of this book is a lifelong learning technique based on a proposed extension to Hidden Markov Models, called Growing Hidden Markov Models (GHMM). They differ from conventional HMMs in the fact that the model structure and parameters are not fixed, but evolve continuously, as more observations are available. This is possible thanks to the integration of a topology learning network [Jockusch and Ritter, 1999] – which is used to learn the structure – and an incremental parameter learning algorithm inspired on the work of Neal and Hinton [1998] into a fully unsupervised learning technique.

Even if they have been formulated in the context of motion prediction, GHMMs are intended to be applicable to all the cases in which standard HMMs are used as a discrete approximation to model a process having a continuous state space.

In order to apply GHMMs to the motion prediction problem, we have taken a different approach from other HMM based techniques, which – as explained above – are based in the concept of typical trajectories. Instead, we have chosen to model motion in terms of the object's intended destination (*ie* its goal), which, in our approach, becomes part of an extended state vector, along with the other state variables such as pose and velocity.

By using together GHMMs and the extended state, we have developed a novel motion prediction technique, which has the following differences with respect to other HMM-based approaches:

- The model's parameters and structure are estimated using an incremental algorithm, which is a necessary condition for true lifelong learning.
- The learning algorithm is fully unsupervised, and it is defined in terms of intuitive parameters, which do not require any prior knowledge about the environment or the number of possible motion patterns.
- The learned HMM structure does not consist of unconnected chains representing 'typical trajectories' corresponding to geometrically similar observation sequences. Instead, motion is modeled in terms of goals. This not only enables richer motion pattern representations, but also has the advantage of being an explicit – although rough – model of the object's intentions.
- In spite of being richer than in other approaches, the learned HMM structure is still simple enough to enable the use of exact inference in real time, even for large environments and high-dimensional spaces: in most cases, the number of edges grows linearly with the number of discrete states in the model.

1.4 Overview of the Rest of This Book

Part One: Background

Chapter 2. This chapter introduces the basic mathematical notions of probabilities and inference which are fundamental for this work.

Part Two: State of the art

Chapter 3. This chapter is a review of the literature of learning-based motion models and their application to prediction. It proposes a classification of motion models, analyzes existing techniques, and discusses outstanding issues on intentional motion prediction, making emphasis in problems related to discrete space-state representations such as Hidden Markov Models.

Chapter 4. Since most of our work is based on Hidden Markov Models, we have consecrated one chapter to this probabilistic tool. In it, we discuss common inference tasks like filtering, smoothing and prediction; and present an overview of existing parameter and structure learning algorithms.

Part Three: Proposed approach

Chapter 5. This chapter introduces Growing Hidden Markov Models, our proposed HMM extension for continuous parameter and structure learning. It presents the probabilistic model and introduces the concept of a topological map before presenting the details of our learning algorithm and an overview of its properties.

Chapter 6. This chapter explains the application of GHMMs to predict the motion of vehicles and pedestrians. It explains the reasons and advantages of using an extended state vector and describes how it is integrated into the model. Finally, it compares our proposed approach against other HMM-based techniques in the literature.

Part Four: Experiments and Applications

Chapter 7. This chapter presents the different experimental platforms used in this book as well as the respective data sets.

Chapter 8 This chapter presents the qualitative and quantitative results that we have obtained from the application of our GHMM based motion prediction technique to the data sets described in the precedent chapter. It also discusses the comparison we have conducted against two other state-of-the-art approaches.

Part Five: Conclusions

Chapter 9. Finally, this chapter summarizes this book and analyzes our contributions. Then, it presents our conclusions and reviews some of the possibilities for future work.

Part I
Background

Chapter 2
Probabilistic Models

- How do you know?
- All those movies had happy endings.
- All?
- Most.
- That cuts down the probability - he told her, smug.

THOMAS PYNCHON
The crying of lot 49

2.1 Overview

Probabilities and Bayesian models – in particular the Bayes filter – will play a rather predominant role in what is to follow, so that it seems advisable to begin with a summary introduction to the subject. This chapter is strongly based on Lebeltel [1999], the interested reader is also invited to consult the works of Stone et al. [1999] and Thrun et al. [2005] for a more thorough explanation of the concepts presented here.

We place ourselves in the framework of the theory of plausible reasoning developed by Jaynes [1995] "probability as logic"; an extension of classical logic which introduces probabilities as a way to perform inference. One of the key aspects of this theory is its interpretation of probabilities as describing the degree of certainty that it is possible to have about a particular phenomenon given the available knowledge. This *subjective* approach is a significant departure from the classic *frequentist* point of view, which sees probabilities as an inherent property of the phenomenon.

The basis of Jaynes work is a theorem proposed by Cox [1946], who first stated a number of desiderata defining the concept of plausibility for a reasonable agent. From there, he showed that the only way to manipulate this notion while staying true to the original desiderata is by using the mathematical

A.D. Vasquez: Incre. Learn. for Motion Predict. of Pedest. & Vehicles, STAR 64, pp. 11–24.
springerlink.com

concept of probabilities. This result is crucial for a "scientifically respectable theory of inference"[1]: although estimating probabilities is subjective because it depends on the agent's knowledge, given the same knowledge, two different agents should estimate the same probabilities if they are rational.

2.2 From Logic to Probabilities

2.2.1 Logic Propositions

We will work with classic logic propositions which may be either true or false. These propositions will be denoted using capital calligraphic characters. Propositions may be manipulated using Boolean operators.

- *Logical AND.* We will denote the logical product, or conjunction, by listing propositions together, separated by spaces, for example $\mathcal{A}\mathcal{B}$ means "both $\mathcal{A}$ *and* $\mathcal{B}$ are true". Sometimes, for the sake of clarity, we will use a comma to indicate conjunction (*eg* "$\mathcal{A}, \mathcal{B}$"). Finally, as a notational convenience, we will use colon-separated sub indices to denote the conjunction of variables with consecutive indices(*eg* $\mathcal{A}_{1:t} = \mathcal{A}_1, \cdots, \mathcal{A}_t$).
- *Logical OR.* Logical disjunction will always be denoted by using a plus sign, hence $\mathcal{A} + \mathcal{B}$ means "at least one of $\mathcal{A}$ *or* $\mathcal{B}$ is true".
- *Negation.* Negation will be denoted by the symbol $\neg$, therefore $\neg\mathcal{A}$ means "$\mathcal{A}$ is not true".

2.2.2 Probability of a Proposition

Sometimes, we do not know if a particular proposition $\mathcal{A}$ is true or not, but we still have reasons (*eg* prior knowledge or evidence) that make us believe that one of those values is more likely than the other. We express this using the notation $P(\mathcal{A} \mid \pi)$ which is read "the conditional probability of $\mathcal{A}$ *given* our former knowledge π". Since – under the subjective interpretation – probabilities are always estimated on the basis of former knowledge, it does not make sense to simply write $P(\mathcal{A})$ or "probability of $\mathcal{A}$". That said, we will systematically omit the specification of former knowledge π, but it should be noted that this is only a notational shortcut.

2.2.2.1 Quantitative Rules for Propositions

Quantitative rules are mathematical identities used to manipulate proposition probabilities in order to perform inference, in a similar fashion to what we do with Boolean operators.

[1] [Jaynes, 1995, p.39].

The product rule

The product rule relates the probability of the logical product $\mathcal{A}\ \mathcal{B}$ to the individual probabilities of $\mathcal{A}$ and $\mathcal{B}$:

$$P(\mathcal{A}\ \mathcal{B}) = P(\mathcal{A})P(\mathcal{B} \mid \mathcal{A}) = P(\mathcal{B})P(\mathcal{A} \mid \mathcal{B}) \tag{2.1}$$

The normalization rule

The normalization rule expresses the relationship between the probability of a proposition and that of its negation.

$$P(\mathcal{A}) + P(\neg\mathcal{A}) = 1 \tag{2.2}$$

The sum rule

Just as in classical logic it is possible to construct every possible logical expression using only negation and conjunction, two identities – the product and the normalization rules – are used to derive all possible probabilistic computations. One of the rules which is possible to obtain from (2.1) and (2.2) is the sum rule:

$$P(\mathcal{A} + \mathcal{B} \mid \mathcal{C}) = P(\mathcal{A} \mid \mathcal{C}) + P(\mathcal{B} \mid \mathcal{C}) - P(\mathcal{A}\ \mathcal{B} \mid \mathcal{C}) \tag{2.3}$$

2.2.3 Variables

Until now, we have expressed concepts in terms of logical propositions but, most often, we want to reason in terms of variables (also called random variables) which represent the relevant features of a given problem. A discrete variable V has an associated domain $D_V = \{v_1, \cdots, v_K\}$, which is the set of K values which this variable may take. The variable may only have a single value at the same time, and the value should be in D_V.

Let us suppose that we want to model the result of throwing a die, representing the output by a discrete variable V having a domain $D_V = \{1, \cdots, 6\}$. We may associate a logic proposition to every possible output:

$$\begin{aligned} \mathcal{V}_1 &\equiv [V = 1] \\ &\vdots \\ \mathcal{V}_6 &\equiv [V = 6] \end{aligned}$$

These propositions are both *exhaustive* and *mutually exclusive.*

$$\mathcal{V}_1 + \cdots + \mathcal{V}_6 = 1 \,(\text{exhaustive})$$
$$\mathcal{V}_i\mathcal{V}_j = 0 \;\forall i,j \;\big|\; i \neq j \,(\text{mutually exclusive})$$

We may now use the obtained propositions to define probabilities, for example, the probability of obtaining a 6 may be written $P(\mathcal{V}_6)$, alternatively, we may explicitly enclose the proposition with square brackets $P([V = 6])$.

If we have two variables V^1 and V^2 whose respective domains are $D_{V^1} = \{v_1^1, \cdots, v_N^1\}$ and $D_{V^2} = \{v_1^2, \cdots, v_M^2\}$, we may describe $N \times M$ mutually exclusive elementary propositions:

$$\begin{gathered}
\mathcal{V}_1^1\mathcal{V}_1^2, \mathcal{V}_1^1\mathcal{V}_2^2, \cdots, \mathcal{V}_1^1\mathcal{V}_N^2 \\
\mathcal{V}_2^1\mathcal{V}_1^2, \mathcal{V}_2^1\mathcal{V}_2^2, \cdots, \mathcal{V}_2^1\mathcal{V}_N^2 \\
\vdots \\
\mathcal{V}_M^1\mathcal{V}_1^2, \mathcal{V}_M^1\mathcal{V}_2^2, \cdots, \mathcal{V}_M^1\mathcal{V}_N^2
\end{gathered}$$

Each of these $N \times M$ propositions may be interpreted as follows:

$$\mathcal{V}_i^1\mathcal{V}_j^2 \equiv [V^1 = v_i^1][V^2 = v_j^2]$$

"The value of V^1 is v_i^1 *and* the value of V^2 is v_j^2"

Since these propositions are both exhaustive and mutually exclusive, the conjunction $V^1\ V^2$ is itself a new random variable.

2.2.3.1 Quantitative Rules for Discrete Variables

Whereas there are some differences between variables with discrete and real domains, we will treat them individually. For discrete variables, the product rule is:

$$\begin{aligned}
&\forall\, i \in D_A,\; j \in D_B \\
&P([A = i]\ [B = j]) = P([A = i])P([B = j] \mid [A = i]) \\
&\qquad\qquad\qquad\quad\;\; = P([B = j])P([A = i] \mid [B = j])
\end{aligned} \tag{2.4}$$

In this book, we will use the variable name alone (*ie* without square brackets) to denote the *entire* probability distribution. Hence the above expression may be noted with the following shortcut:

$$P(A\ B) = P(A)P(B \mid A) = P(B)P(A \mid B) \tag{2.5}$$

In general, the probability of the conjunction of two or more variables is called a Joint Probability Distribution (JPD). The factorization of the JPD into simpler probabilities is called a *decomposition.*

The normalization rule may at first look somewhat different from (2.2). Intuitively, both of them may be regarded as stating that the sum of the probabilities for all the possible cases is one:

$$\sum_{i \in D_A} P([A = i]) = 1 \tag{2.6}$$

Again, this may be written:

$$\sum_{A} P(A) = 1 \tag{2.7}$$

2.2.3.2 Quantitative Rules for Continuous Variables

Of course, many problems are defined in terms of variables having a continuous domain. In this case, probabilities are defined by assuming that there is a Probability Density Function (PDF) $g(A)$ which defines the probability of an interval:

$$P([a < A \leq b]) = \int_{A=a}^{b} g(A)dA \tag{2.8}$$

For continuous variables, the product rule is the same that in (2.5):

$$P(A\ B) = P(A)P(B \mid A) = P(B)P(A \mid B) \tag{2.9}$$

And the normalization rule becomes:

$$\int_{-\infty}^{\infty} g(A)dA = 1 \tag{2.10}$$

2.2.3.3 Other Useful Identities

A very useful identity involving variables is the *theorem of total probability*, also known as the *marginalization rule*:

$$\sum_{B} P(A\ B) = \sum_{B} P(B)P(A \mid B) = P(A) \qquad \text{(discrete case)} \tag{2.11}$$

$$\int_{-\infty}^{\infty} P(A\ B)dB = \int_{-\infty}^{\infty} P(B)P(A \mid B)dB = P(A) \qquad \text{(continuous case)} \tag{2.12}$$

Equally important is the so called *Bayes rule*, which establishes the relationship between the "forward probability" $P(A \mid B)$ and its "inverse" $P(B \mid A)$, this is particularly useful in inverse problems, where the parameters of a model need to be estimated from data:

$$P(A \mid B) = \frac{P(B \mid A)P(A)}{P(B)} = \frac{P(B \mid A)P(A)}{\sum_A P(B \mid A)P(A)} \quad \text{(discrete case)} \tag{2.13}$$

$$P(A \mid B) = \frac{P(B \mid A)P(A)}{P(B)} = \frac{P(B \mid A)P(A)}{\int_{-\infty}^{\infty} P(B \mid A)P(A)dA} \quad \text{(continuous case)} \tag{2.14}$$

It is important to note that, since the denominator does not depend on A

$$P(A \mid B) \propto P(B \mid A)P(A) \tag{2.15}$$

and the proportionality coefficient may be deduced from the marginalization rule.

The Bayes rule may be seen as a mechanism to compute (*ie* infer) the probability of a variable A, whose value we ignore, at the light of a variable B, whose value is known. The known variable is often called the *evidence*, the term $P(A \mid B)$ is called the posterior probability or simply the *posterior* and the term $P(A)$ is called the prior probability or simply the *prior*. Finally, the term $P(B \mid A)$ is often called *data likelihood*[2].

2.2.4 JPD Decomposition and Conditional Independence

The use of the quantitative rules is not limited to the case of two variables: since the product of two or more random variables is itself a random variable, it is possible to apply the rules to an arbitrary number of variables.

The dependencies between different variables are formally denoted by their joint probability, and the way it is decomposed into simpler probability distributions by applying the product rule. For example, for the case of three variables $\{V^1, V^2, V^3\}$ there are 13 possible decompositions of the JPD:

[2] There is a lot more to say about Bayes rule. For example, the importance of priors and their use to describe former knowledge is a source of debate between frequentists and Bayesians (*eg* subjectivists). However, such a discussion is beyond the scope of this book.

$$
\begin{aligned}
P(V^1, V^2, V^3) &= P(V^1)P(V^2\ V^3 \mid\ V^1) \\
&= P(V^2)P(V^1\ V^3 \mid\ V^2) \\
&= P(V^3)P(V^1\ V^2 \mid V^3) \\
&= P(V^1\ V^2)P(V^3 \mid V^1\ V^2) \\
&= P(V^1\ V^3)P(V^2 \mid V^1\ V^3) \\
&= P(V^2\ V^3)P(V^1 \mid V^2\ V^3) \\
&= P(V^1)P(V^2 \mid V^1)P(V^3 \mid V^1\ V^2) \\
&= P(V^1)P(V^3 \mid V^1)P(V^2 \mid V^1\ V^3) \\
&= P(V^2)P(V^1 \mid V^2)P(V^3 \mid V^1\ V^2) \\
&= P(V^2)P(V^3 \mid V^2)P(V^1 \mid V^2\ V^3) \\
&= P(V^3)P(V^1 \mid V^3)P(V^2 \mid V^1\ V^3) \\
&= P(V^3)P(V^2 \mid V^3)P(V^1 \mid V^2\ V^3)
\end{aligned}
$$

These 13 factorizations are equivalent from a mathematical point of view, however, by choosing a particular decomposition, we express our knowledge about the structure of variable dependence, as related to a particular problem or application.

After choosing a decomposition, further simplification is possible on the basis of *conditional independence* assumptions, which apply whenever we consider that the variable V^1 does not provide additional information about a variable V^2 if a third variable V^3 is known:

$$
\begin{aligned}
P(V^1\ V^2\ V^3) &= P(V^1)P(V^2 \mid V^1)P(V^3 \mid V^1\ V^2) \\
&= P(V^1)P(V^2 \mid V^1)P(V^3 \mid V^1)
\end{aligned}
$$

These assumptions are based on knowledge about the variables' semantics, they may also be introduced to simplify, even if they are not necessarily true. Anyway, conditional independence assumptions are critical for tractability since they permit to reduce the dimensionality and, therefore, the complexity of the distributions which conform the JPD.

2.2.5 Inference

The rules we have described so far may be used to put together a model which describes some phenomenon or process; such a probabilistic model is specified by enumerating the variables – and their domains – which are considered to be relevant to the phenomenon being modeled, as well as the corresponding JPD decomposition.

The main application of probabilistic models is inference: finding out the values of unknown variables on the basis of known variables (*ie* evidence) through the application of the Bayes rule. More formally, inference consists of finding the probability distribution of a set of variables ε_r (*ie* researched variables) on the basis of another set of variables ε_k, whose value we know (*ie* evidence), eventually disregarding the values of a third group of variables ε_i (*ie* ignored variables). We define a probabilistic question as any expression having the form:

$$P(X^k \ \cdots \ X^l \mid [X^m = x^m] \ \cdots \ [X^n = x^n]) \tag{2.16}$$

Where $\varepsilon_r = \{X^k, \cdots, X^l\} \neq \varnothing$, $\varepsilon_k = \{X^m, \cdots, X^n\}$ and $\varepsilon_i = \{X^o, \cdots, X^p\}$ is the set of variables that do not belong to neither ε_r nor ε_k.

Knowing the JPD, it is possible to answer any probabilistic question of the form (2.16) by first applying the product rule (2.1):

$$P(X^k \ \cdots \ X^l \mid [X^m = x^m] \ \cdots \ [X^n = x^n]) = \frac{P(X^k \ \cdots \ X^l \ [X^m = x^m] \ \cdots \ [X^n = x^n])}{P([X^m = x^m] \ \cdots \ [X^n = x^n])} \tag{2.17}$$

And then expressing the numerator and the denominator in terms of the JPD by applying the marginalization rule (2.12):

$$\cdots = \frac{\sum\limits_{X^o \cdots X^p} P(X^k \ \cdots \ X^l \ [X^m = x^m] \ \cdots \ [X^n = x^n] \ X^o \ \cdots \ X^p)}{\sum\limits_{\substack{X^k \cdots X^l \\ X^o \cdots X^p}} P(X^k \ \cdots \ X^l \ [X^m = x^m] \ \cdots \ [X^n = x^n] \ X^o \ \cdots \ X^p)} \tag{2.18}$$

As Cooper [1990] has demonstrated, Bayesian inference is NP-hard; this explains the importance of conditional independence assumptions, which render the problem tractable by allowing further simplifications. Another way to cope with complexity is by using specially tailored inference algorithms which exploit specific features of a model; it is also possible to trade-off accuracy against complexity by using approximate inference algorithms. Some examples of these algorithms will be given in § 2.3.

2.2.6 *Parametric Forms*

Up to this moment, we have not really explained how the simpler probability distributions that compose the JPD are defined; indeed, a complete probabilistic model description should specify how these probabilities are represented. Throughout this book, these probabilities will be chosen amongst a few elementary distributions, which are defined in terms of a number of parameters, and hence are also known as *parametric forms.*

2.2.6.1 Uniform Distribution

The uniform distribution (noted $\mathbf{U}_V$) represents the fact that all the values in the domain of a variable V are equiprobable. In general, it is only well defined for discrete variables:

$$P([V = v_i]) = \mathbf{U}_V(v_i) = \frac{1}{|D_V|}, \forall v_i \in D_V \tag{2.19}$$

The only parameter of this distribution is the cardinality of the variable's domain $|D_V|$.

2.2.6.2 Conditional Probability Table

Conditional probabilities on discrete variables may be represented as a Conditional Probability Table (CPT), which lists the probabilities that the variables on the left side of the bar may take depending on the evidence variables. for example, if a, b and c are all binary variables (*ie* $D_a, D_b, D_c \in \{0, 1\}$), the probability of a given b and c is represented with a $2 \times 2 \times 2$ table t whose individual elements will be represented with comma-separated sub indices:

$$P([A = i] \mid [B = j]\ [C = k]) = T_{i,j,k}$$

In general a conditional probability distribution consisting of N variables $\{V^1, \cdots, V^N\}$ will correspond to an N dimensional CPT with $|D_{V^1}| \times \cdots \times |D_{V^N}|$ elements or parameters.

2.2.6.3 Gaussian Distribution

For continuous variables we will often use the multivariate Gaussian (or Normal) probability distribution, which has the following form[3]:

$$P([V = v_i]) = \mathbf{G}(v_i;\ \mu,\ \Sigma) \tag{2.20}$$

$$= |2\pi\Sigma|^{-1/2} \exp\left[-\frac{1}{2}(V - \mu)^T \Sigma^{-1} (V - \mu)\right] \tag{2.21}$$

where μ is the mean vector, Σ is a $D \times D$ positive semidefinite matrix called *covariance matrix* and $v_i, \mu \in \mathbb{R}^D$. The mean and covariance matrix[4] constitute the Gaussian's parameters.

2.2.7 Learning

Besides the definition of the probabilistic model itself, it is necessary to assign values to the parameters (*eg* mean value and covariance) of every elementary probability distribution in the decomposition. Although this may be done

[3] Strictly speaking, the expression presented here does not denote a probability since, as it may be seen in eq. (2.8), continuous probability distributions are only defined for intervals. However, it is common in the literature to refer to PDFs as probabilities [*eg* Thrun et al., 2005], which is what we will do in this document.

[4] In the scalar (unidimensional) case, the covariance is called simply variance.

manually, it is also possible to learn (*ie* estimate) the parameters' values from data.

For discrete variables, in its simplest form, learning consists in computing the frequencies at which different variable's values occur in data, this is done by counting. So, if we have only one variable V with domain D_V:

$$P([V = v_i]) = \frac{C([V = v_i])}{\sum_V C(V)} \tag{2.22}$$

where $C(\mathcal{A})$ stands for the number of cases in data where proposition $\mathcal{A}$ is true. This is known as a *maximum likelihood* estimate, since it maximizes data likelihood without taking into account any prior on the distribution of V. A problem with this approach is that, when a learning case does not appear in data, the corresponding probability becomes zero. This disregards the – very frequent – situation in which data is not exhaustive, *ie* it does not contain an example of every possible value of V. The problem may be solved by using Dirichlet priors, that is, pre assigning counts α_i to every value v_i in D_V:

$$P([V = v_i]) = \frac{C([V = v_i])) + \alpha_i}{\sum_{v_j \in D_V} C([V = v_j]) + \alpha_j} \tag{2.23}$$

the particular case in which all $\alpha_i = 1$, becomes equal to a uniform in the absence of data; it is known as Laplace's law of succession:

$$P([V = v_i]) = \frac{C([V = v_i]) + 1}{\sum_V C(V) + 1} \tag{2.24}$$

Laplace's law may be generalized to learn an arbitrary conditional probability:

$$\begin{aligned} P([V^1 = v^1] \ \cdots \ [V^k = v^k] \mid [V^l = v^l] \ \cdots \ [V^m = v_m]) \\ = \frac{C([V^1 = v^1] \ \cdots \ [V^k = v^k] \ [V^l = v^l] \ \cdots \ [V^m = v_m]) + 1}{\sum\limits_{V^1,\cdots,V^k} C(V^1 \ \cdots \ V^k \ [V^l = v^l] \ \cdots \ [V^m = v_m]) + 1} \end{aligned} \tag{2.25}$$

In many real situations, data is not available for some of the variables in the model, which are then said to be *hidden*. The standard solution is the use of the Expectation-Maximization (EM) algorithm [Dempster et al., 1977], which starts with some initial (often random) estimate of the parameters and then uses inference to compute the expected number of counts – called Expected Sufficient Statistics (ESS) – which are then treated as though they were observed and used to re-estimate the parameters. This two steps are iterated until some convergence criterion is met.

A much more challenging learning problem is conditional structure learning, which aims to estimate from data the best JPD decomposition and conditional independence assumptions. The general approach is to evaluate different decompositions using a scoring function, selecting the one which obtains the best score.

Since the literature about learning probabilistic models is huge, we will not further discuss general methods in this chapter[5]. Instead, in chapter 4 we will do a more thorough review of the learning algorithms that apply to the particular probabilistic model upon which we have based our work: *Hidden Markov Models.*

2.3 The Bayes Filter

We will conclude this chapter by introducing the Bayes filter, which is used to study dynamic systems and constitutes the basic probabilistic framework for chapters 4 and 5.

The objective of the Bayes filter is to find a probabilistic estimate of the current state of a dynamic system – which is assumed to be "hidden", *ie* not directly observable – given a sequence of observations gathered every time-step up to the present moment.

The main advantage of using a probabilistic framework such as the Bayes filter is that it allows to take into account the different sources of uncertainty that participate in the process, such as:

- The limited precision of the sensors used to obtain observations.
- The variability of observations due to unknown factors (observation noise).
- The incompleteness of the model.

2.3.1 Probabilistic Model

2.3.1.1 Variables

The Bayes filter works with two types of variables[6]:

- S_t, the state of the system at time t. The exact meaning of this variable depends on the particular application, in general, it may be seen as the set of system features which are relevant to the problem and have an influence in the future. Since this work deals mostly with object motion, we will frequently assume that the state is the object's pose, although the specific variables that constitute the state may vary with the context and include additional features such as velocity. This will be explicitly stated in the text.

[5] See [Heckerman, 1995, Murphy, 2002] for good introductions to the subject.

[6] Here, we are providing the definition of Stone et al. [1999], which is different from the one given by Thrun et al. [2005] in that the later includes a third set of variables describing actions or controls.

- O_t, the observation gathered at time t. Observations provide indirect indications about the state of the system. In this book, we will assume that observations come from sensors which "observe" a given environment, such as laser scanners and video trackers.

The Bayes filter is an abstract model which does not make any assumption about the nature (*ie* discrete or continuous) of the state and observation variables. Such assumptions are made by concrete specializations, such as the Kalman Filter (continuous state and observations) or Hidden Markov models (discrete state, discrete/continuous observations).

2.3.1.2 Decomposition

A Bayes filter defines a joint probability distribution on $O_{1:T}$ and $S_{1:T}$ on the basis of two conditional independence assumptions:

1. Individual observations O_t are independent of all other variables given the current state S_t:

$$P(O_t \mid O_{1:t-1}\ S_{1:t}) = P(O_t \mid S_t) \tag{2.26}$$

   In general $P(O_t \mid S_t)$ is called *observation probability* or *sensor model.* It models the relationship between states and sensor readings, taking into account factors such as accuracy and sensor noise.
2. The current state depends only on the previous one, knowledge about former states do not provide any further information. This is also known as the order one *Markov hypothesis*:

$$P(S_t \mid S_{1:t-1}) = \begin{cases} P(S_1) & \textit{for } t = 1 \\ P(S_t \mid S_{t-1}) & \textit{otherwise} \end{cases} \tag{2.27}$$

   The probability $P(S_t \mid S_{t-1})$ is called the *transition probability* or *system model.* The probability $P(S_0)$ that describes the initial state in the absence of any observations is called the *state prior.*

   These assumptions lead to the following decomposition of the Bayes filter:

$$P(S_{1:T}\ O_{1:T}) = P(S_1)P(O_1 \mid S_1)\prod_{t=1}^{T} P(S_t \mid S_{t-1})P(O_t \mid S_t) \tag{2.28}$$

2.3.2 Parametric Forms

The Bayes filter does not define any parametric forms and, consequently, no parameter identification mechanism. This is left out to specializations.

2.3.3 Inference

One of the main uses of Bayes filters is to answer the probabilistic question $P(S_{t+H} \mid O_{1:t})$; what is the state probability distribution for time $t + H$, knowing all the observations up to time t? The most common case is filtering ($H = 0$) which estimates the current state. However, it is also frequent to perform prediction ($H > 0$) or smoothing ($H < 0$).

The Bayes filter has a very useful property that largely contributes to its interest: filtering may be efficiently computed by incorporating the last observation O_t into the last state estimate using the following formula:

$$P(S_t \mid O_{1:t}) = \frac{1}{Z} P(O_t \mid S_t) \sum_{S_{t-1}} [P(S_t \mid S_{t-1}) P(S_{t-1} \mid O_{1:t-1})] \tag{2.29}$$

where, by convention, Z is a normalization constant which ensures that probabilities sum to one over all possible values for S_t.

If we define recursively $P(S_{t-1}) = P(S_{t-1} \mid O_{1:t-1})$, it is possible to describe a Bayes filter in terms of only three variables: S_{t-1}, S_t and O_t, leading to the following decomposition:

$$P(S_{t-1}\ S_t\ O_t) = P(S_{t-1}) P(S_t \mid S_{t-1}) P(O_t \mid S_t) \tag{2.30}$$

Where the state posterior of the previous time step is used as the prior for the current time and is often called *belief state* Thrun et al. [2005].

Under this formulation, the Bayes filter is described in terms of a local model, which describes the state's evolution during a single time step. For notational convenience, in the rest of this book we are only going to describe such local models, noting that they always describe a single time step of the global model (2.29).

2.3.4 Specializations of the Bayes Filter

Different choices of variable domains and parametric forms for the Bayes filter give place to different filtering techniques. For example, Kalman Filters [Kalman, 1960] make the following hypotheses:

- State and observation variables are both continuous.
- The three probabilities in the decomposition are Gaussians.
- Both the evolution of the system state and the dependence between the system and the observation variables are described by linear functions.

Other popular specializations of the Bayes filter are Hidden Markov Models – which will be reviewed in detail in chapter 4 – and particle filters [Arulampalam et al., 2002], which represent probabilities using discrete samples, and allow for fast approximated inference. Particle filters will not be further discussed in this book.

2.4 Discussion

In this chapter we introduced the basic mathematical tools needed for understanding the following chapters. We have briefly presented the use of probabilities as a modeling and inference tool which allows the explicit representation of uncertainties.

We discussed the basic rules used to manipulate probabilities – *ie* the product and normalization rules – as well as some very useful derived rules – *ie* the sum and marginalization rules. We explained the components of a probabilistic model: the variables, joint probability decomposition and the parametric forms. We presented the Bayes rule and the fundamental role it plays in inference. We described the use and principles behind learning algorithms, in particular the importance of counting as a means of estimating probabilities. We have also introduced the Bayes filter and explained the variables that constitute it: state and observations.

Part II
State of the Art

Chapter 3
Intentional Motion Prediction

> Here is how it works: first you decide to treat the object whose behavior is to be predicted as a rational agent; then you figure out what beliefs that agent ought to have, given its place in the world and its purpose. Then you figure out what desires it ought to have, on the same considerations, and finally you predict that this rational agent will act to further its goals in the light of its beliefs. A little practical reasoning from the chosen set of beliefs and desires will in most instances yield a decision about what the agent ought to do; that is what you predict the agent will do.
>
> DANIEL DENNETT
> *The Intentional Stance*

3.1 Overview

In the epigraph, Dennett [1987] describes how most humans would try to predict another humans' motion if asked to: if we accept that a moving object is rational and that it is able to move at will – we take the *intentional stance* towards it – then it follows that the best way to predict its motion is to try to emulate its rational process, *eg* "the car is slowing down because it wants to park in the free place at the left". But designing a computer program capable of such emulation in a general fashion is far beyond our current capabilities.

Instead, most intentional motion prediction approaches adopt what Dennet calls the *physical stance*[1]: they try to explain – and predict – the object's

[1] It is important to note that the physical and intentional stances stand in opposite ends of a continuum instead of describing mutually exclusive views.

A.D. Vasquez: Incre. Learn. for Motion Predict. of Pedest. & Vehicles, STAR 64, pp. 27–43.
springerlink.com

behavior in terms of its physical properties and the laws of physics. This implies the use of kinematic and dynamic models describing the objects' motion as the evolution of its state (*eg* its position, velocity and orientation) over time when it is subject to a control (*eg* acceleration).

However, this approach faces at least three problems:

- The current object's state is not directly known, instead, it is necessary to estimate it from observations gathered through sensors which have a limited precision and are subject to noise.
- Kinematic and dynamic models are inevitably incomplete, because, in order to be general, they deliberately disregard a number of factors that influence motion (*eg* road conditions, wind, etc.).
- Although objects following intentional motion certainly obey physical laws, they are able to dynamically modify their motion by applying a control (*eg* steering angle, braking) hence, in order to predict motion, it is necessary to predict the control that the object will apply.

The first problem may be addressed by using continuous Bayes filters such as the Kalman Filter [Kalman, 1960] and its extensions [*eg* Rao, 1996, Julier and Uhlmann, 1997]. However, this does not solve the other two issues and, when applied to pedestrian and vehicles, approaches based on Kalman filters [*eg* Han and Veloso, 1997, Rosales and Sclaroff, 1998, R.Madhavan and Schlenoff, 2003] produce predictions which are sound only during a short time interval – also known as their *time horizon* – or which are only applicable to highly structured environments. The same happens to approaches based on regression analysis [*eg* Elnagar and Gupta, 1998, Yu and Su, 2003, Liu et al., 2004], which have the additional drawback of producing deterministic predictions, thus being unable to represent the uncertainty that is inherent to the prediction process.

More recently, another family of approaches has emerged, based on an idea which is closer to the intentional stance: they assume that, in a given environment, objects do not move at random but often engage in typical behaviors or motion patterns. Therefore, if these motion patterns are known, it is possible to use them as motion models in order to predict motion.

This solves the third problem, since now it is only necessary to identify the motion pattern that the object is following, instead of predicting the applied control; indeed, it is not longer necessary to model controls explicitly In addition, most of these approaches address the second problem by means of a probabilistic representation, where the inherent incompleteness of the model is reflected as uncertainty.

We will hereafter refer to this last family of approaches as "pattern based motion models"[2]. This chapter is a review of the related literature, it is

[2] As opposed to "kinematic" or "dynamic" motion models.

structured according to a sub-classification of these approaches in three categories, depending on how they represent motion patterns:

1. *Trajectory Prototypes.* As their name suggests, these approaches model motion patterns using a trajectory prototype (*ie* typical example) for every motion pattern.
2. *Discrete state-space Models.* This models are based on *Markov chains*, where time progresses as discrete steps and, at every time step, the objects' state evolves by going from one state to another according to a given transition probability.
3. *Other representations.* Here we will discuss those approaches that do not fit into the other categories.

For every category of approaches, we are particularly interested in studying how they answer three questions:

1. *Representation.* How are motion patterns described mathematically?
2. *Learning.* How are the model's parameters estimated from data?
3. *Prediction.* When an object is moving, what is the process that transforms observations into predictions?[3]

Throughout this chapter, when needed, we will base our discussion on the imaginary office environment depicted in fig. 3.1, where some pedestrian trajectories have been collected and are used as input for the different techniques discussed here. In this example, trajectory data is assumed to consist of noisy measurements of a person's pose, evenly sampled in time.

After presenting existing approaches, we will conclude with a discussion of relevant global trends and issues in the field.

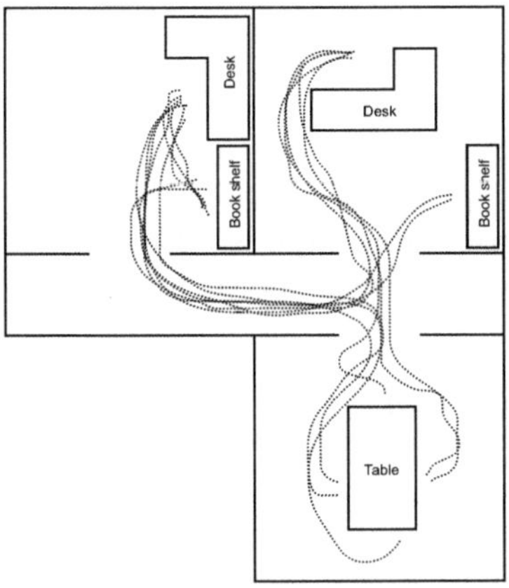

Fig. 3.1 Raw trajectory data of people in an office environment.

[3] We have decided to include in this review several approaches that do not specifically address the prediction question, but apply learning algorithms to build motion models, which, at least in theory, could be used to predict motion.

3.2 A Note on Semantics

Until now, we have indistinctly used the terms "behavior" and "motion pattern", this coincides with most of the reviewed works, which do not agree on a standard definition of these concepts. Henceforth, we will consider that a behavior describes what an object is doing by taking the intentional stance towards the object, while a pattern is the motion that may be observed when the object is involved in a particular behavior. We will illustrate this distinction with an example:

Let us imagine two persons in a hall, the first one is reading a bulletin board, while the second one is waiting for someone. From the persons' perspective these are two completely different behaviors, however, they produce very similar motion patterns: the persons do not move during a long time.

This distinction is important. At first sight it seems that, since the patterns are similar, they may be described by the same motion model. But this will lead us to dismiss a crucial difference: the "reading behavior" only happens in the bulletin board's neighborhood, while the "waiting behavior" may occur anywhere in the hall. As a consequence, our model will fail to represent the fact that it is much more likely to observe a motionless person near the bulletin board than elsewhere.

This is related to the notion of state. Most of the techniques that we have reviewed assume that the state variables represent only physical properties of an object such as its pose or velocity. Nonetheless, other representations are possible, as in [Oliver et al., 2000], where states have higher level semantics and objects may be in a "waiting" or a "fleeing" state and, therefore, are closer to modeling behavior.

3.3 Trajectory Prototypes

Approaches in this category work by grouping similar trajectories in clusters (fig. 3.2) – which correspond to typical motion patterns – then, for every cluster, a single trajectory prototype is obtained and used to represent the entire cluster and, in consequence, the corresponding motion pattern.

3.3.1 Representation

Typical trajectories are often represented as a sequence of points in continuous state-space. Most approaches do not model the time variable explicitly and assume that the points in the sequence are regularly spaced in time.

Sometimes, a measure of the "width" of the cluster is also included in the representation (fig. 3.3). For example, Makris and Ellis [2001] construct a "path model" by including a deterministic measure of the width for every point in the trajectory, a similar measure has been used by Junejo et al. [2004]. Vasquez and Fraichard [2004] propose a probabilistic model in which

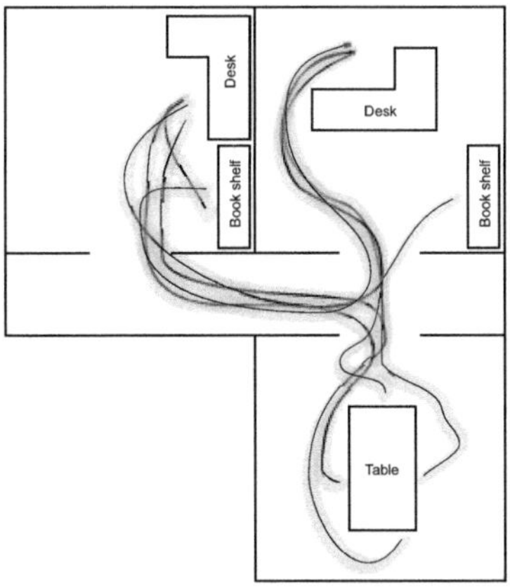

Fig. 3.2 Trajectory prototypes for the office environment.

the width of the trajectory is represented as the variance of the distance between the trajectories that belong to the same cluster. Another probabilistic model of width has been proposed by Bennewitz et al. [2002], they model every point of the trajectory prototype as a Gaussian and assume that all such Gaussians have the same covariance matrix.

3.3.2 Learning

Trajectory prototypes are obtained using classical clustering algorithms [see Kaufman and Rousseeuw, 1989, Jain et al., 1999]. Three basic problems should be solved by a learning algorithm: a) determining the number of clusters to be found; b) finding clusters; and c) building trajectory prototypes from clusters, including, if necessary, the width representation.

There are many different classification schemes of clustering (*eg* hard vs. soft, deterministic vs. stochastic, see Jain et al. [1999]) here, we will distinguish between *model based clustering algorithms*, which work by deriving a set of prototype vectors; and *pairwise clustering algorithms*, which partition data into disjoint sets or clusters by performing pairwise comparisons between data elements. The interest of this classification is that algorithms of different categories provide different solutions to the three problems that we have mentioned above.

3.3.2.1 Model Based Algorithms

Model based algorithms have the advantage that finding the trajectory prototypes is implicit to the clustering process. They also have interesting theoretical properties since they are built by optimizing a global measure (*eg* data likelihood).

However, they need to know *a priori* the number of clusters – thus, of motion patterns – to be found, meaning that this number should be estimated somehow. Some approaches just ignore this problem, for example, Hu et al. [2004b] assume that the number of clusters is known and use a custom self organizing network to cluster trajectories. Others propose greedy search techniques, like, Bennewitz et al. [2002] which estimate the number of models by using the EM algorithm [Dempster et al., 1977] to cluster data using an initial guess of the number of clusters, and then adding or deleting clusters in order to maximize data likelihood, rerunning EM after every addition.

An issue with model based clustering is that most algorithms assume that their input consists of vectors of equal length, which is problematic because trajectory lengths may vary greatly. The most frequent solution [*eg* Hu et al., 2004b] is to resample trajectories in order to normalize their length. Unfortunately, this goes against the assumption that the points in trajectory prototypes are evenly spaced in time, and temporal information is lost in the process. An alternative has been proposed in [Bennewitz et al., 2002], it consists in normalizing all the trajectories in the data set to the length of the longest trajectory by padding the end of shorter trajectories with copies of its last position; this may be interpreted as an object which stops moving, but is still detected for a while.

3.3.2.2 Pairwise Clustering

Pairwise clustering algorithms are based on the use of a dissimilarity measure (*eg* Euclidean distance) to compare two data elements (*ie* trajectories). The value of the measure is high if the elements are very different and zero if they are equal. Data elements are processed in pairs and the dissimilarity measure is used to decide if elements belong to the same cluster. The clustering process produces groups of data elements, but no cluster representation. This is equivalent to adding a label to every element, indicating the group to whom it belongs. When compared with model based clustering algorithms, these approaches have the advantage that they are able to determine the number of clusters. On the other hand, since they are based in a local pairwise measure, it is not possible to guarantee that they satisfy a global optimality criterion.

In order to use pairwise clustering to learn motion patterns, it is necessary to choose a clustering algorithm, a dissimilarity measure and to device a mechanism to compute a cluster representation.

A popular pairwise clustering algorithm is Hierarchical Clustering [King, 1967], Makris and Ellis [2001] used it in conjunction with a maximum interpoint distance; while Buzan et al. [2004] used longest common subsequence (LCSS) as similarity measure; and Vasquez and Fraichard [2004], used an extension of the Euclidean distance for trajectories.

The use of graph cut algorithms [Boykov and Kolmogorov, 2004] has also been explored by Junejo et al. [2004], using the Hausdorff distance as dissimilarity measure. A custom pairwise algorithm has been proposed by

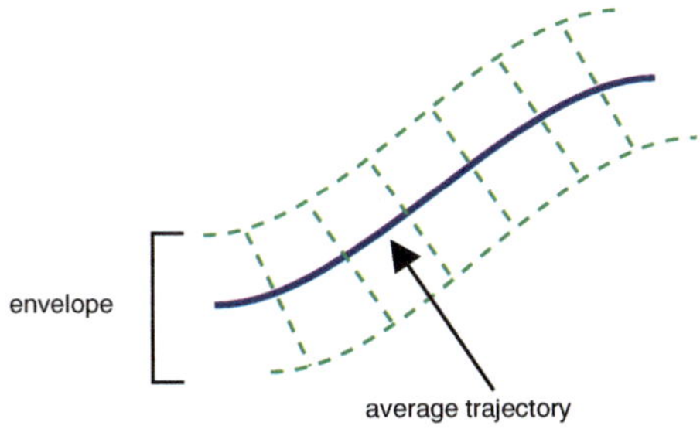

Fig. 3.3 Example of average trajectory and deterministic envelope.

Wang et al. [2006] based on a distance criterion augmented with a measure of confidence for observations, which permits to represent sensor noise and limited precision.

Once that clustering has been performed, the average trajectory for each cluster is computed and used to represent the cluster. Then, the cluster width may be computed either probabilistically – by computing a cluster variance [Vasquez and Fraichard, 2004] – or deterministically, by defining a cluster "envelope" [*eg* Junejo et al., 2004] which is obtained from the trajectories in the cluster that are farthest away from the average (fig. 3.3).

3.3.3 Prediction

Prediction using trajectory prototypes consists basically in finding the cluster which best corresponds to a partially observed trajectory – often using the same dissimilarity measure that has been used to cluster – and taking that cluster's representation as a prediction of future motion. A variant has been proposed by Vasquez and Fraichard [2004] which returns a probability distribution on all the clusters according to their similarity.

While these approaches have the advantage of being very coherent in the long-term – which, in theory, allows for very long time-horizons – they suffer from two important drawbacks: only trajectories that have been observed may be predicted; and, most important, they have a strictly deterministic representation of time, the predicted state for an object at time t corresponds to the t-th element of the trajectory prototype, which is too restrictive, given that velocities often vary between objects even if they are engaged in the same behavior. The problem becomes even more severe when trajectories are resampled to normalize their length. As a consequence, uncertainties may be handled only at the trajectory level, in other words, if the motion pattern that the object is following is known, these algorithms will provide a deterministic prediction of the state.

A solution to the last problem is to transform trajectory prototypes into state-space models, it will be discussed in the following section.

3.4 Discrete State-Space Models

As their name indicates, these approaches are based on the use of a discrete model as an approximate analysis tool for continuous motion. The basis of discrete state-space approaches are Markov chains, which represent time as a discrete variable and model motion in terms of a finite number of discrete states. Transitions between states depend on a transition probability, which follows the Markov assumption, namely that, knowing the present state, the future and past states are independent. Markov chains may be represented by a directed graph, whose edges are labeled with the probability of going from one state to another in a single time step (fig. 3.4).

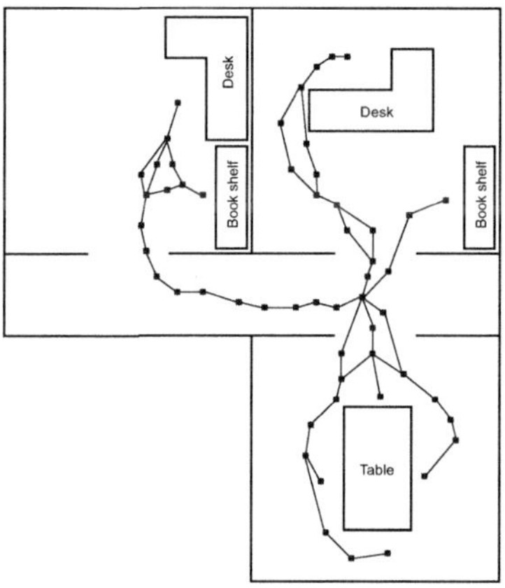

Fig. 3.4 Discrete state-space model for the office environment (transition directions are not depicted).

Markov chains, however, fail to model the uncertainty related to sensors, and only a few motion prediction techniques use them (*eg* Tadokoro et al. [1995], J. Rittscher and Stein [2003]). Observation uncertainty is specifically addressed by using discrete Bayes filters (cf. 2.3), which augment Markov chains by assuming that the state is not directly observable and should be inferred from sensor readings or *observations* to which they are related by an observation probability. One of the most popular discrete Bayes filters are Hidden Markov Models (HMM), which we have briefly described in the introduction and will be discussed in detail in chapter 4.

Most of the techniques presented in this section are based on the HMM framework. However, they are also applicable to other discrete state-space models unless noted otherwise.

3.4.1 Representation

As explained in chapter 1, there are mainly two ways to represent multiple behaviors using HMMs:

One approach (*eg* Makris and Ellis [2002], Bennewitz et al. [2005]) is to represent all the motion patterns using a single Hidden Markov Model, where different patterns share no states and correspond to non connected components in the graph representation of the model; the other approach (*eg* Walter et al. [1999]) is to use a different HMM per motion pattern and model the belief as a mixture, where the mixing variable represents the probability of a particular HMM being the "good" model. In fact, the former solution may be seen as a special case of the latter, where the mixing variable is always uniform.

Closely related to motion pattern representation is the concept of structure, which may be regarded as defining the transition graph by choosing a number of states and selecting which state transitions are considered to be 'valid' or possible.

3.4.2 Learning

Learning state-space based motion pattern representations may be decomposed in two subtasks: a) learning the structure – *ie* the topology – of the underlying transition graph; and b) learning the parameters, *ie* the transition probabilities associated with every edge of the graph, and the observation probabilities describing how observations are related to states. Sometimes also a prior on the state is also learned, but often it is assumed to be uniform [*eg* Walter et al., 1999, Bennewitz et al., 2005].

3.4.2.1 Structure Learning

Despite the existence of several structure learning algorithms for state-space models [*eg* Stolcke and Omohundro, 1994, Friedman, 1997, Brand, 1998], they are seldom used in the context of motion pattern learning. A notable exception is the work of Brand and Kettnaker [2000] which uses an entropic structure estimator that will be described in more detail in §4.6. Instead, structure is either fixed *a priori* or estimated using custom mechanisms.

One of such mechanisms is the use of trajectory prototype learning techniques in order to identify both the number and the form of motion patterns, and then transforming those prototypes into a state-space model. A good example is the approach proposed by Bennewitz et al. [2005], they obtain trajectory prototypes using EM as explained in §3.3, then they transform those prototypes into an HMM whose parameters are not learned, but manually assigned on the basis of domain knowledge and a number of assumptions about how objects move. Both [Koller-Meier and Van Gool, 2001] and

[Makris and Ellis, 2002] combine trajectory prototypes and state-space models in a similar way, but the later also incorporates parameter learning into the algorithm.

Minnen and Wren [2004] present an algorithm which performs hierarchical decomposition of data to find structure at many levels of detail. The decomposition is performed by constructing a binary tree where each node contains two HMMs. The HMMs are used to classify data into two subclasses, corresponding to the node's sub trees. Classified data is sent to the respective sub tree and the bifurcation process continues until no further decomposition is possible.

An approach that integrates knowledge about the environment has been proposed by Liao et al. [2003], they use information on the static features of the environment to compute a Voronoi graph, which is then used to define the structure of a custom state-space model.

3.4.2.2 Parameter Learning

Having determined a structure, parameter learning may be performed by standard means using the Baum-Welch algorithm [Baum et al., 1970], which is a specialization of the well known Expectation Maximization algorithm [Dempster et al., 1977]. This is the case of Makris and Ellis [2002] and Liao et al. [2003]. Other authors prefer to determine parameters' values using custom procedures. Bennewitz et al. [2005] model observation probabilities as Gaussians, all having the same covariance; while transition probabilities are computed analytically under the hypothesis that objects speeds follow a Gaussian distribution.

3.4.3 Prediction

In state-space models, motion is predicted using Bayesian inference. Since the model structure is determinant in the complexity of inference, it is possible to find approaches that use either exact inference (*eg* Brand and Kettnaker [2000]), or approximate inference methods, such as particle filters [Arulampalam et al., 2002] (*eg* Walter et al. [1999], Koller-Meier and Van Gool [2001]).

3.4.4 Other State-Space Models

Several other extensions of Hidden Markov Models have also been used for representing intentional motion, here we provide a short overview of them.

Pentland and Liu [1999] have proposed Markov Dynamic Models (MDM) which, conceptually are exactly like HMMs, except that the nodes of the graph represent controls (*eg* acceleration) and not the object's pose. Instead of state observations, MDM track the object with a Kalman filter and use the innovation (*ie* prediction error) of the filter as observations. Although they have shown that MDMs outperform HMMs in recognition tasks, it is difficult to use them to predict state since there is no pose representation.

The work of Kruse et al. [1996, 1997], Kruse and Wahl [1998] is also worth mentioning, they have developed a technique where motion is represented by line segments where the object is supposed to move at uniform speed and transitions do not occur between states, but between line segments.

Bui et al. [2002] have proposed Abstract Hidden Markov Models (AHMM), which are a hierarchical extension of HMMs that allow to represent motion at different levels of detail (*eg* metric, room, building, etc.) and integrate the concept of policy (*ie* plan), which may be regarded as the equivalent of a motion pattern. The usage of AHMMs to model motion in big scale environments has been demonstrated by Liu et al. [2004], with a three-layer model representing, from top to bottom, goals, transportation modes and the physical state of the object. The approach is able to learn the goal and transportation mode structures using custom tailored algorithms, but the low-level physical structure is given *a priori* in the form of a graph. Another AHMM-based approach has been proposed by Osentoski et al. [2004], using EM to learn the model's parameters. One of the advantages of AHMMs over HMMs is that they are able to provide good mid-term predictions even if they are wrong at the long-term, for example, they may accurately predict that a person is going towards the room's door even if they wrongly predict that the final destination is the kitchen. Nevertheless, unsupervised learning with AHMMs is challenging, mainly because of the difficulty of automatically finding a good, semantically sound, hierarchical structure.

All the approaches we have mentioned until now make the hypothesis that, when there is more than one object in the environment, their motion is independent, thus ignoring object interaction. This problem has been studied by Brand et al. [1997] by means of an HMM extension called Coupled Hidden Markov Models (CHMM), which represents the joint state of all objects. The main drawback of this approach is that the number of possible interactions grows exponentially with the number of interacting objects. This has motivated the work of Gong and Xiang [2003], Xiang and Gong [2006] which have proposed Dynamically-multi-linked Hidden Markov Models (DML-HMM), which are based on the idea that interactions should be modeled only when there is a high probability that they actually take place.

A different direction has been taken by Oliver et al. [2000], they use CHMMs to model interaction behaviors between pairs of objects (*eg* "approach", "meet" and "continue together") without modeling the environment at all. Behaviors are defined prior to learning and trained on labeled data. This approach has been very successful in explaining human behavior; however, since there is no pose representation, it is not applicable to motion prediction. The same can be said about more recent motion pattern recognition approaches, such as the one proposed by Hoogs and Perera [2008], that models behavior with a dynamic Bayesian network containing semantic states obtained from a predefined ontology.

3.5 Other Approaches

3.5.1 Neural Network Based Approaches

Johnson and Hogg [1995], and Sumpter and Bulpitt [2000] have proposed approaches based on multilayer neural networks: the first layer is used to discretize the space by using a Self-organizing network to perform Vector Quantization (VQ) so that every unit (*ie* neuron) in this layer corresponds to a discrete state. In order to obtain the pattern representation, both approaches use an additional layer of "leaky neurons", one for every discrete state. These leaky neurons turn "on" when an object passes through the corresponding state and their output decays slowly afterwards. As a result, the leaky neuron layer keeps a "trace" of the trajectory after the object's motion has ended; this trace (*ie* the activation values of the leaky neurons) is used to represent the learned pattern. Finally, different patterns are classified using a third layer of neurons, which performs VQ on the whole output vector of the second layer, *ie* the values of all the leaky neurons.

Hu et al. [2004a] suppress the leaky network layer by adding lateral connections between the units in the state layer in order to account for the temporal relationships between them.

A drawback which is common to all these approaches is that they assume that the number of motion patterns to be represented is known *a priori*, which implies a fair amount of knowledge about the environment. An approach that does not share this problem, has been proposed by Stauffer and Grimson [2000]: instead of the leaky neuron layer, they represent motion patterns using a co-occurrence matrix C where every element $c_{i,j}$ corresponds to the probability that an object passes through states i and j given that it is engaged in a motion pattern.

A different neural network approach has been explored by Chang and Song [1997], they train a multilayer perceptron obtain predictions on a single motion pattern, obtaining good results, however, they do not address the case of having multiple motion patterns.

Neural network based approaches, have a performance which is comparable with that of discrete state-space models, as may be seen in the results obtained by Hu et al. [2004a]. On the downside, these approaches do not represent uncertainty explicitely and, in most cases, rely in ad-hoc procedures to produce probabilistic output.

3.5.2 Goal Oriented Approaches

Dee and Hogg [2004] proposed an approach which takes the intentional stance towards the mobile object. They model motion as being motivated by the desire of the object to move to explicit places (*ie* goals) in the environment. The originality of the approach resides in the fact that the object's beliefs are taken into account by modeling the world from the agent's perspective (*eg* may the object see the wall?). This makes it, to our knowledge, the only approach of this type to model the object's perception. From there, motion is modeled on the basis of a cost model by assuming that the object will always take the cheapest path to its goal unless something happens. However, although this is interesting for scene interpretation (*ie* if the object did not take the cheapest path, then something happened), it is not as useful for motion prediction because of its overly simple and deterministic motion model. Also, the approach assumes that the general structure of the environment is known *a priori.*

Foka and Trahanias [2002] have also modeled motion in terms of goals, they predict motion in terms of a simple geometric method which does not take the environment into account. A similar approach has been proposed in Bruce and Gordon [2004] where the geometric method is complemented with a stochastic motion model.

3.5.3 Other Fields

The problem of building models of temporal processes has also been studied in other disciplines. For example, Dixon et al. [2004] have proposed a programming-by-demonstration framework that represents motion using HMMs that are learned with a model merging algorithm very similar in spirit to the one proposed by Stolcke and Omohundro [1993]. Another programming-by-demonstration framework has been proposed by Calinon and Billard [2007] it represent trajectories as Gaussian Mixture Models and uses Gaussian Mixture Regression for inference. The paper proposes two incremental parameter learning strategies, while the model structure (*ie* number of Gaussians) is fixed *a priori.*

Finally, we would like to mention the work of Kulic et al. [2008] in the context of whole body motion. They propose an approach based on a generalization of HMMs called Factorial Hidden Markov Models. The approach assumes a chain-like structure and fixed number of discrete states and proposes an algorithm to estimate the model parameters and to instantiate new motion models. The approach builds an HMM out of every input observation sequence and uses a symmetric version of the Kullback-Leibler divergence as the basis for a Hierarchical Agglomerative clustering algorithm where new motion models are inserted according to a threshold.

3.6 Discussion

This chapter provided a general review of the area of pattern based motion models. Now we would like to draw some general conclusions about existing techniques, in view of achieving our main objective: building a lifelong motion learning and prediction technique to improve autonomous navigation.

3.6.1 General Issues

Here, we discuss some issues that are general to all pattern based motion models. When relevant, we will discuss how different approaches deal with them, focusing mainly in the difference between trajectory prototypes and state-space models.

3.6.1.1 Incremental Learning

First of all, a lifelong learning approach should be incremental. This means that it should verify three properties [Langley, 1995]:

- It should process data elements one by one.
- It should not reprocess any previous data element.
- It should retain only one main knowledge structure in memory.

The main reason for this requirement is efficiency. Using a batch algorithm would mean storing the complete history of observation sequences, and processing it after every trajectory has observed. This is ineffective and not well suited for continuous processing in real-time.

3.6.1.2 Representing Uncertainty

In order to plan its trajectory, an autonomous vehicle needs to know in advance the state of the different obstacles that are present in the environment. At the same time, since uncertainty is inherent to prediction, a deterministic output would have little chances to be true and would be of limited utility. The alternative is to explicitly represent the uncertainty by using a probabilistic representation.

Since state-based approaches are built on probabilistic models they are ideal candidates for motion prediction approaches. Trajectory prototypes, on the other hand, are deterministic models. Even when they are complemented with a variance as in [Bennewitz et al., 2002, Vasquez and Fraichard, 2004] so that they are able to model spatial uncertainty, they still fail to model the temporal uncertainty which is related to the fact that in real environments, objects not only do not move at the same speed but, most of the time, their speeds vary as they move.

3.6.1.3 Discovering New Motion Patterns

An incremental pattern learning approach should be able to identify when an object is involved in an unknown behavior and to create a new pattern model in consequence. This is, in general, difficult to implement for state-space models, since it involves the capacity to dynamically insert new states and connect them with existing ones, which is equivalent to incremental structure learning, that, as we will show in chapter 4, is a difficult problem.

Approaches based on trajectory prototypes seem to be better suited for the task. Even if most of the current approaches are designed to work off-line, it seems feasible to implement an incremental trajectory clustering approach. The same may be said about goal-oriented approaches, such as [Foka and Trahanias, 2002] and [Dee and Hogg, 2004].

3.6.1.4 Representing Interaction between Moving Objects

A good part of human behavior involves interacting with other humans: two persons who find themselves face to face in a corridor will move in order to avoid each other, a man may go faster in order to join a friend who is walking just in front of him, etc.

But dealing with interactions between objects is complex. Modeling interactions with a state-space model implies representing the joint state of all the objects that may be simultaneously present in a scene. If we have N discrete states and M moving objects, this means learning and storing the probabilities of $O(N^M)$ possible interactions, which quickly becomes intractable.

Therefore, discrete states modeled by approaches such as [Brand et al., 1997, Oliver et al., 2000] do not represent places in the space, but a limited number – *ie* two or three – of high-level intentional states such as *avoiding* or *meeting*. This means that these techniques may not be directly used to predict the physical state of an object. However, it would be interesting to study how they may be integrated with other techniques in order to obtain such a prediction. Besides some general remarks in chapter 9, this problem is not addressed in this book.

3.6.1.5 On-Board Sensors

Practically all the techniques that we have reviewed in this chapter assume that there is some kind of global sensor providing information about the motion of all the objects that move in the environment. Learning and modeling intentional motion using on-board sensors is a much more difficult problem that is beyond the scope of this book.

3.6.2 State-Space Model Issues

As we have mentioned before, approaches based on trajectory prototypes are not well suited to produce probabilistic state predictions. As a matter of fact, our previous work on motion prediction [Vasquez and Fraichard, 2004] was based on this kind of approaches. We were able to obtain a probability distribution representing our belief about the motion pattern that the object is actually following. But we also wanted to have a probabilistic representation of the state even when the "right" trajectory is known with certainty. The idea was to take into account the limited precision of our model as well as the variability of the real trajectories which are represented by the same prototype. This ultimately lead us to discrete state-space models and to the approach presented in this book. This section discusses some issues which are exclusive to this family of approaches.

Even when the semantics of discrete states are clear, and they are assumed to represent places in the environment, it is possible to construct different space representations which fall into one of two wide categories[4]: a) *metric* representations, where states represent points in the space, generally represented as a set of coordinates; and *topological* representations, where the states represent only a few interesting places which often possess a high-level semantic connotation (*eg* kitchen, corridor) and which are connected together by some notion of adjacency which is frequently represented with a graph (fig.3.5).

Even though topological representations may be seen as a particular case of metric representations – after all, a metric representation also encodes "interesting places" and their spatial relations – it is frequent to talk of a topological representation when the state variable has a low cardinality. For navigation purposes, metric representations provide more precise and useful information. At the same time, they imply a higher number of states, thus, higher computational complexity.

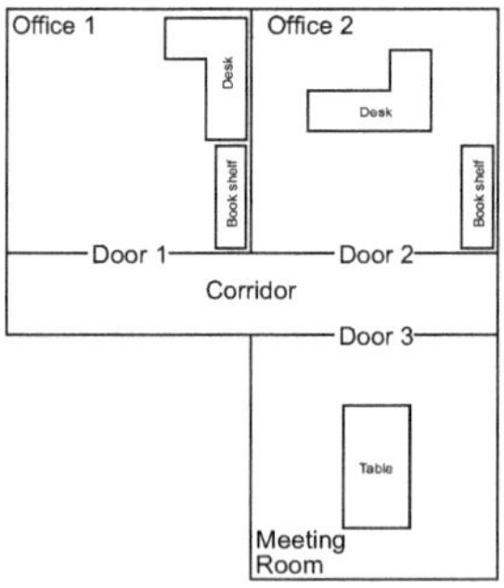

(a) Environment

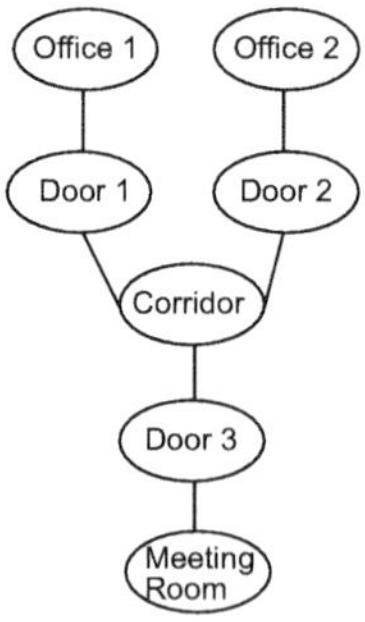

(b) topological representation

Fig. 3.5 A topological representation of the office environment

[4] See [Filliat, 2001] for an extended analysis of these alternatives.

A related subject is that of automatically finding a "good" space decomposition. Answering this involves finding a compromise between precision and computational complexity, because both augment with the number of states. Thrun et al. [2005] distinguish two types of approaches to decompose the space: a) *static* decompositions such as grids, which partition the environment without taking into account how objects move in it; and b) *dynamic* approaches, which adapt to observed motion, performing finer decomposition on areas where objects move and coarser decomposition elsewhere. Although static approaches are easier to implement, they are wasteful on computational resources because they partition the space even in places where no activity takes place. Moreover, due to the "curse of dimensionality"[5], the problem worsens with the number of dimensions in the space (*eg* position, velocity, size). An example of the difference on these approaches is shown in fig.3.6.

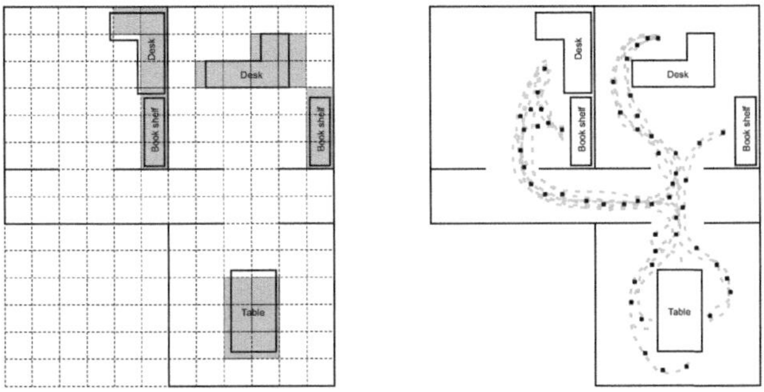

(a) static (168 elements) (b) dynamic (54 elements)

Fig. 3.6 Comparison between static and dynamic decompositions. Notice how the dynamic decomposition produces a better representation of the data (grey dots) using less than a third of the discrete elements of the grid approach.

[5] This term, first employed by Bellman [1961], refers to the exponential growth of the hypervolume of a space, when new dimensions are added; and how it demands a similar increase on the number discrete elements needed to represent the space.

Chapter 4
Hidden Markov Models

Carr. - ... your name isn't Jack, it's Tristan.
Tzara. - No, it isn't, it's Jack.
Carr. - You have always told me it was Tristan. I have introduced you to everyone as Tristan. You answer to the name of Tristan. Your notoriety at the Meieri Bar is firmly associated with the name Tristan. It is perfectly absurd saying your name isn't Tristan.
Tzara. - Well, my name is Tristan in the Meieri Bar and Jack in the library, and the ticket was issued in the library.

Sir Tom Stoppard
Travesties

4.1 Overview

This book build upon the use of Hidden Markov Models as motion models, which, as we have seen in chapter 3, are probably the most popular technique for pattern based motion prediction. This chapter provides the reader with a broad introduction to this probabilistic framework. Sections 4.2 through 4.4 present what may be considered as the "classic" theory on HMMs. Readers already familiar with HMMs may safely skip these sections and proceed directly to section 4.5 and the rest of this chapter, where we discuss structure learning, a difficult problem for which no standard algorithm exists yet, despite the existence of several approaches in the literature.

4.2 Probabilistic Model

Hidden Markov Models (HMMs) are a specialization of the Bayes filter for discrete state variables, they do not constrain the form of observation variables,

A.D. Vasquez: Incre. Learn. for Motion Predict. of Pedest. & Vehicles, STAR 64, pp. 45–68.
springerlink.com

which may be either discrete [Rabiner, 1990] or continuous [Juang et al., 1986]. Although most of the theory for both types of observation variables is the same, some implementation differences exist (*eg* parametric forms, learning algorithm). Here, we have privileged the discussion of continuous observation HMMs because they are better suited to our particular problem.

4.2.1 Variables

Since we are going to use a local model (see §2.3.3), an HMM may be defined in terms of three variables:

- S_t, S_{t-1}, the current and previous states, which are discrete variables with value $S_t, S_{t-1} \in \{1, \cdots, N\}$ for a fixed N.
- O_t, the observation variable, which is a multidimensional vector in $\mathbb{R}^M$.

4.2.2 Decomposition

The decomposition is the same than for the Bayes filter:

$$P(S_{t-1}\ S_t\ O_t) = P(S_{t-1}) P(S_t \mid S_{t-1}) P(O_t \mid S_t) \tag{4.1}$$

Where the state prior is computed recursively:

$$P(S_{t-1}) = P(S_{t-1} \mid O_{1:t-1}) \tag{4.2}$$

4.2.3 Parametric Forms

Hidden Markov Models make an additional conditional independence assumption with respect to the standard Bayes filter: both the observation and the transition probabilities are considered to be *stationary*, that is, independent of time. Thus the parametric forms of the three probabilities in the JPD are always the same, irrespectively of the value of the time variable:

- $P([S_1 = i]) = \pi_i$. The state prior is stored as a a vector $\pi = \{\pi_1, \cdots, \pi_N\}$ where each element contains the prior probability for the corresponding discrete state.
- $P([S_t = j] \mid [S_{t-1} = i]) = a_{i,j}$. Transition probabilities are represented with a $N \times N$ *transition matrix* A where each element $a_{i,j}$ represents the probability of reaching state j in the next time step given that the system is currently in state i.
- $P(O_t \mid [S_t = i]) = \mathbf{G}(O_t;\ \mu_i,\ \sigma_i)$. The observation probability density function is represented by a Gaussian distribution for every

state. The set of all the Gaussians' parameters are denoted as $b = \{(\mu_1, \sigma_1^2), \cdots, (\mu_N, \sigma_N^2)\}$[1].

We denote the whole set of parameters for an HMM by $\lambda = \{\pi, A, b\}$.

4.2.4 *Example: The Broken Air Conditioning System*

We will illustrate the process of defining a continuous observation HMM with a simple example[2]. Let us suppose that we have an air conditioning system which may be in one out of two configurations: "fresh" and "cold"; due to an internal malfunction, the system switches randomly between them once a day according to the following rules:

1. If the current configuration is "fresh", it may change to "cold" with probability 0.8, or stay as it is with probability 0.2.
2. If the current configuration is "cold", it may change to "fresh" with probability 0.9, or stay as it is with probability 0.1.

These rules are presented as a graph in fig. 4.1:

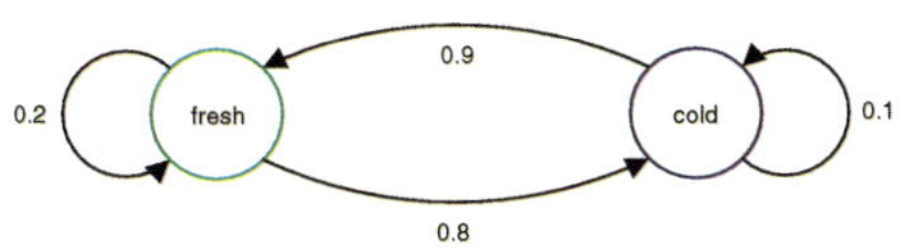

Fig. 4.1 Transition graph of the air conditioning example.

Supposing that – according to its configuration – the machine regulated the room temperature to two different and well defined temperatures, it would be trivial to infer directly the state by measuring their values; unfortunately, this is not the case, and the results of applying one configuration or the other are rather approximative. They may be described probabilistically as follows: when the machine is in the "fresh" configuration, the temperature has a mean value of 20° Celsius and a variance of 25. For the "cold" configuration, the mean temperature is 15° Celsius and the variance is 16.

Representing this system as an HMM is straightforward. First, we define the domains of the state (*ie* configuration) and observation (*ie* room temperature) variables:

---

[1] An extension of this idea known as mixtures of Gaussian densities, is also frequently used to approximate arbitrarily shaped observation probabilities, see [Juang et al., 1986] for further explanation.

[2] This has been inspired on an example of discrete observation HMMs which appears in [Manning and Schutze, 1999].

$$D_{S_t} = \{1, 2\} \quad \text{where 1 stands for "fresh" and 2 for "cold"}$$
$$D_{O_t} = \mathbb{R}$$

then, we fix the parameters' values. First we assign the state prior by assuming that, the first day, both configurations are equiprobable:

$$\pi = [0.5, 0.5]$$

Transition probabilities are filled according to the specification:

$$A = \begin{bmatrix} 0.2 & 0.8 \\ 0.9 & 0.1 \end{bmatrix}$$

Similarly for the parameters of the observation probability:

$$b = \{(20, 25), (15, 16)\}$$

We have now a complete HMM definition. Of course, we would like to use it to answer questions about the behavior of the system, which is the subject of the following section.

4.3 Inference

Inference tasks on HMMs may be classified in: a) on-line inference, which takes place as the system evolves, every time that a new observation is available; and b) off-line inference, whose input consists of complete observation sequences and, therefore, is performed after the observed phenomenon has finished. The distinction is important because most off-line inference tasks are based on a dynamic programming technique called the forward-backward algorithm which will be introduced in §4.3.2.

4.3.1 On-Line Inference

On-line inference tasks are particularly useful for applications like navigation, where it is necessary to take decisions while objects are moving. The key to on-line inference lies in maintaining the belief state, from there, inference may be performed in terms of the local model by using the belief state as a prior and applying Bayesian inference.

4.3.1.1 State Filtering

As explained in §2.3.3, state filtering is one of the most common probabilistic questions for Bayes filters – thus, for HMMs – the name comes from the fact that its purpose is to "filter out" observation noise in order to estimate the system state. It is done using expression (4.3), which is conceptually divided in two steps: a) *prediction* which projects the current belief state one time step into the future; and b) *update*, which improves the estimation by integrating the last obtained observation:

$$P(S_t \mid O_{1:t}) = \underbrace{\frac{1}{Z} P(O_t \mid S_t)}_{update} \underbrace{\sum_{S_{t-1}} [P(S_t \mid S_{t-1}) P(S_{t-1} \mid O_{1:t-1})]}_{prediction} \tag{4.3}$$

Prediction is the most computationally expensive step, typically, this operation involves iterating through the N possible transitions for every state, hence, its time complexity is $O(N^2)$. However, as we will show later, the complexity of the prediction step may be reduced by imposing constraints on the structure of the transition matrix.

4.3.1.2 Prediction

Predicting the future state consists basically in, taking the belief state and then propagating it ahead in time, it may be regarded as filtering without the update step (eq. (4.4)). Given that it is necessary to iterate through every time step up to H – also known as the *time horizon* – the algorithm's complexity is $O(HN^2)$.

$$P(S_{t+H} \mid O_{1:t}) = \sum_{S_{t+H-1}} [P(S_{t+H} \mid S_{t+H-1}) P(S_{t+H-1} \mid O_{1:t})] \tag{4.4}$$

4.3.1.3 Example

Continuing with our example, let us suppose that we want to perform filtering to estimate the state of our AC machine from temperature readings. The first day, we get a temperature reading of 23.05°, so, we compute for $S_1 = 1$:

$$\begin{aligned} P([S_1 = 1] \mid [O_1 = 23.05]) &= P([O_1 = 23.05] \mid [S_1 = 1]) P([S_1 = 1]) \\ &= \frac{1}{Z_1} \mathbf{G}(23.05;\ 20,\ 25) \pi_1 \\ &= \frac{1}{Z_1} [0.066][0.5] \\ &= \frac{0.033}{Z_1} \end{aligned}$$

this is a special situation because $t = 1$ and we do not have previous observations of the system's state. Thus, instead of applying the prediction step, we just use the state prior $P([S_1 = 1])$.

Now, we compute for $S_1 = 2$:

$$\begin{aligned} P([S_1 = 2] \mid [O_1 = 23.05]) &= P([O_1 = 23.05] \mid [S_1 = 2])P([S_1 = 2]) \\ &= \frac{1}{Z_1}\mathbf{G}(23.05;\ 15,\ 16)\pi_2 \\ &= \frac{1}{Z_1}[0.013][0.5] \\ &= \frac{0.007}{Z_1} \end{aligned}$$

substituting the value of the normalizing variable $Z_1 = 0.033 + 0.007 = 0.040$, we obtain the following probabilities.

$$\begin{aligned} P([S_1 = 1] \mid [O_1 = 23.05]) = 0.83 \\ P([S_1 = 2] \mid [O_1 = 23.05]) = 0.17 \end{aligned}$$

These probabilities constitute the belief state for $t = 1$, which may be written as a vector:

$$\hat{S}_1 = [0.83, 0.17]$$

The second day, we measure again the temperature, now getting a reading of 17.5°. We proceed in an analogous fashion to estimate the system state for $S_2 = 1$:

$$\begin{aligned} &P([S_2 = 1] \mid [O_1 = 23.05]\ [O_2 = 17.5]) \\ &\quad = P([O_2{=}17.5] \mid [S_2{=}1]) \sum_{S_1} P([S_2{=}1] \mid S_1)P(S_1 \mid [O_1{=}23.05]) \\ &\quad = \frac{1}{Z_2} \times \mathbf{G}(17.5;\ 20,\ 25)[a_{1,1} \times 0.83 + a_{2,1} \times 0.17] \\ &\quad = \frac{1}{Z_2} \times 0.070 \times [0.2 \times 0.83 + 0.9 \times 0.17] \\ &\quad = \frac{0.022}{Z_2} \end{aligned}$$

and for $S_2 = 2$:

$$
\begin{aligned}
P([S_2=2] \mid [O_1=23.05]\ [O_2=17.5]) \\
&= P([O_2{=}17.5] \mid [S_2{=}2]) \sum_{S_1} P([S_2{=}2] \mid S_1) P(S_1 \mid [O_1{=}23.05]) \\
&= \frac{1}{Z_2} \times \mathbf{G}(17.5;\ 15,\ 16)[a_{1,2} \times 0.83 + a_{2,2} \times 0.17] \\
&= \frac{1}{Z_2} \times 0.082 \times [0.8 \times 0.83 + 0.1 \times 0.17] \\
&= \frac{0.056}{Z_2}
\end{aligned}
$$

again, we substitute $Z_2 = 0.022 + 0.056 = 0.078$ to obtain the belief state:

$$\hat{S}_2 = [0.28, 0.72]$$

Notice that every filtering step requires only to process the preceding belief state, instead or iterating through all the sequence of observations from 1 up to t.

Now, we would like to predict the system's state for day 4. First, we predict for $t = 3$:

$$
\begin{aligned}
P([S_3=1] \mid O_{1:2}) &= \sum_{S_2} P([S_3=1] \mid S_2) P(S_2 \mid O_{1:2}) \\
&= a_{1,1} \times 0.28 + a_{2,1} \times 0.72 \\
&= 0.70
\end{aligned}
$$

$$
\begin{aligned}
P([S_3=2] \mid O_{1:2}) &= \sum_{S_2} P([S_3=2] \mid S_2) P(S_2 \mid O_{1:2}) \\
&= a_{1,2} \times 0.28 + a_{2,2} \times 0.72 \\
&= 0.3
\end{aligned}
$$

where the belief state $\hat{S}_2$ constitutes the starting point of the prediction process. Then we compute for $t = 4$:

$$
\begin{aligned}
P([S_4=1] \mid O_{1:2}) &= \sum_{S_3} P([S_4=1] \mid S_3) P(S_3 \mid O_{1:2}) \\
&= a_{1,1} \times 0.7 + a_{2,1} \times 0.3 \\
&= 0.41
\end{aligned}
$$

$$\begin{aligned} P([S_4 = 2] \mid O_{1:2}) &= \sum_{S_3} P([S_4 = 2] \mid S_3) P(S_3 \mid O_{1:2}) \\ &= a_{1,2} \times 0.7 + a_{2,2} \times 0.3 \\ &= 0.59 \end{aligned}$$

The process continues like this for any given time horizon H.

4.3.2 Off-Line Inference

Off-line inference receives its name from the fact that it answers questions about complete observation/state sequences and, in consequence, it can only take place once the last observation has been collected. Examples of offline inference tasks are learning, diagnosis and classification.

Since off-line inference tasks process all the T observations in a sequence, it seems that the cost of inference should grow exponentially with respect to the length of the sequence. Fortunately, this is not the case: thanks to the use of dynamic programming, it is possible to perform exact inference in $O(TN^2)$.

4.3.2.1 The Forward-Backward Algorithm

This algorithm, proposed by Baum et al. [1970], applies dynamic programming techniques to avoid the redundant computations involved in repetitively applying marginalization to perform inference (see §2.2.5). It works by pre-computing two sets of probabilities (*ie* forward and backward) and storing their values in memory. Then, the probabilistic questions are answered by reformulating inference in terms of these variables. Since they are precomputed, most marginalization operations are replaced by memory accesses whose cost is constant; thus significantly reducing the number of operations required to perform inference.

Forward probabilities $\alpha_t(i) = P(O_{1:t}\ [S_t = i] \mid \lambda)$ are computed recursively using the following expression:

$$\alpha_t(i) = \left[\sum_{j=1}^{N} \alpha_t(j) P([S_t = i] \mid [S_{t-1} = j]) \right] P(O_t \mid [S_t = i]) \tag{4.5}$$

This leads to an efficient algorithm (Alg. 1) which computes forward probabilities sequentially, in a similar fashion to state filtering: for every time step, the algorithm iterates through all the states, and, for every state, it iterates through all its predecessors. The whole computation takes $O(TN^2)$ calculations. Fig. 4.2 illustrates the process on our example HMM.

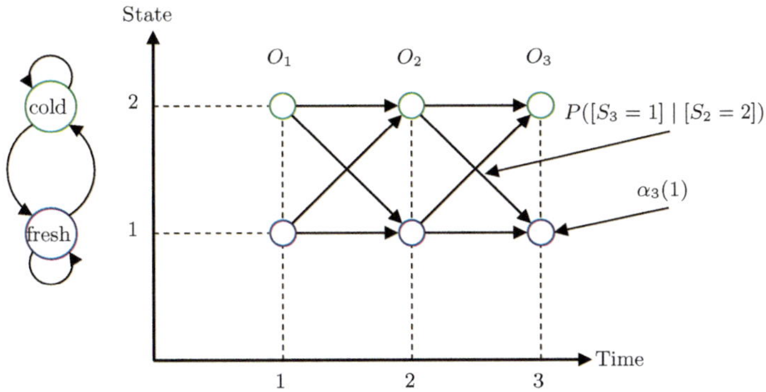

Fig. 4.2 Computing forward probabilities for our example.

The use of forward probabilities to perform inference is nicely illustrated by a question that appears often in HMM related problems: what is the probability of a complete observation sequence given the model parameters?[3]

A naive solution to this problem consists in marginalizing over all the state variables $S_1, \cdots, S_T$, which is equivalent to enumerating all the possible state sequences of length T:

$$P(O_{1:T} \mid \lambda) = \sum_{S_1,\cdots,S_T} P(S_1)P(S_2 \mid S_1)P(O_t \mid S_2)\cdots P(S_T \mid S_{T-1})P(O_T \mid S_T) \tag{4.6}$$

Using this equation, the computation of $P(O_{1:T} \mid \lambda)$ has an $O(N^T)$ complexity, which, is clearly unfeasible in most cases. Fortunately this probability may be computed much more efficiently using forward probabilities by observing that, by definition:

$$\begin{aligned} P(O_{1:T} \mid \lambda) &= \sum_{i=1}^{N} P(O_{1:T}, S_T = i \mid \lambda) \\ &= \sum_{i=1}^{N} \alpha_T(i) \end{aligned} \tag{4.7}$$

Taking into account the cost of computing all $\alpha_t(i)$, we find that $O(TN^2)$ calculations are necessary to compute the observation probability using (4.7). A huge simplification with respect to the $O(N^T)$ computations required by (4.6). Hereafter we will call the probability $P(O_{1:T} \mid \lambda)$ *joint*

[3] If we see the HMM as a generative model – a probabilistic automaton which produces observations – this may be interpreted as the probability that the observation sequence has been produced by the HMM.

Algorithm 1. *Forward_algorithm*$(O_{1:T}, \lambda)$

input : An observation sequence $O_{1:T}$
HMM parameters $\lambda = \{\pi, b, A\}$
output : Forward Probabilities $\alpha_t(i)$

begin
  for $i = 1$ *to* N **do**
    $\alpha_1(i) = P([S_1 = i])P(O_1 \mid [S_1 = i])$
  end
  for $t = 2$ *to* T **do**
    for $j = 1$ *to* N **do**
      $\alpha_t(j) = \left[\sum_{i=1}^{N} \alpha_t(i) P([S_t = j] \mid [S_{t-1} = i])\right] P(O_t \mid [S_t = j])$
    end
  end
end
return *all* $\alpha_t(i)$

Algorithm 2. *Backward_algorithm*$(O_{1:T}, \lambda)$

input : An observation sequence $O_{1:T}$
HMM parameters $\lambda = \{\pi, b, A\}$
output : Backward probabilities $\beta_t(i)$

begin
  for $i = 1$ *to* N **do**
    $\beta_T(i) = 1$
  end
  for $t = T - 1$ *down to* 1 **do**
    for $i = 1$ *to* N **do**
      $\beta_t(i) = \sum_{j=1}^{N} P([S_{t+1} = j] \mid [S_t = i]) P(O_{t+1} \mid [S_{t+1} = j]) \beta_{t+1}(j)$
    end
  end
end
return *all* $\beta_t(i)$

observation probability and, due to its frequent use, we will denote it with the shortand P_O.

Backward probabilities $\beta_t(i) = P(O_{t+1:T} \mid [S_t = i]\ \lambda)$, are also computed recursively:

$$\beta_t(i) = \sum_{j=1}^{N} P([S_{t+1} = j] \mid [S_t = i]) P(O_{t+1} \mid [S_{t+1} = j]) \beta_{t+1}(j) \qquad (4.8)$$

however, there is an important difference with respect to forward probabilities, backward probabilities depend on the *next* time step, hence, they should be computed in inverse order, starting from the last observation (which explains their name). This computation implies a problem: while, in the case of forward probabilities, the state prior is used to initialize the process, nothing similar may be done for $t = T$. The standard algorithm (alg. 2) solves this by arbitrarily setting $\beta_T(i) = 1$.

Examples of the use of backward probabilities to perform inference are *smoothing* (see next section) and parameter learning (§4.4).

4.3.2.2 Smoothing

In most situations, it is possible to obtain a better state estimation if the complete observation sequence is available. This is called *smoothing* and, as we will see, is very important for learning (§4.4).

Smoothing is computed with the forward-backward probabilities:

$$P([S_t = i] \mid O_{1:T}) = \frac{1}{P_O}\alpha_t(i)\beta_t(i) \tag{4.9}$$

Closely related to smoothing and also useful for learning is the probability of a state transition given a sequence of observations:

$$P([S_{t-1} = i]\,[S_t = j] \mid O_{1:T}) = \frac{\alpha_{t-1}(i)P([S_t = j] \mid [S_{t-1} = j])P(O_t \mid [S_t = j])\beta_t(j)}{P_O} \tag{4.10}$$

4.3.2.3 Viterbi Decoding

As a consequence of the state being hidden, a single sequence of observations may correspond to many different state sequences. A question that is often posed to HMMs is "what is the sequence of states which most likely corresponds to a sequence of observations?", which is also called the *best explanation* of the observation sequence. It is answered by maximizing the joint state probability given the observations $\arg\max_{S_{1:T}} P(S_{1:T} \mid O_{1:T})$, which may be computed very efficiently with the *Viterbi algorithm* (alg. 3) [Viterbi, 1967, Forney, 1973].

Essentially, the Viterbi algorithm is identical to the forward algorithm except that the summation in eq. (4.5) is replaced by a maximum operation, according to the following recursion:

$$\delta_t(j) = \max_i \left[\delta_{t-1}(i)P([S_t = j] \mid [S_{t-1} = i])\right] P(O_t \mid [S_t = j]) \tag{4.11}$$

Where $\delta_t(j)$ represents the maximum likelihood of observing the partial observation sequence $O_{1:t}$ and being in state j at time t. The algorithm also stores, for every time t and state j, the preceding state $\psi_t(j)$ which has lead to j with maximum probability:

$$\psi_t(j) = \arg\max_i \{\delta_{t-1} P([S_t = j] \mid [S_{t-1} = i])\} \tag{4.12}$$

When the values for the last time step have been computed, the algorithm computes the best path by finding the state which maximizes $S_T^* = \arg\max_j \psi_T(j)$, and then backtracking according to the predecessor array: $S_{T-1}^* = \psi_T(S_T^*)$ and so on.

Algorithm 3. $Viterbi(O_{1:T}, \lambda)$

---

input : An observation sequence $O_{1:T}$
HMM parameters $\lambda = \{\pi, b, A\}$
output: Most likely state sequence $S_{1:T}^*$

begin
  for $i = 1$ *to* N **do**
    $\delta_1(i) = P([S_1 = i])P(O_1 \mid [S_1 = i])$
    $\psi_1(i) = 0$
  end
  for $t = 2$ *to* T **do**
    for $j = 1$ *to* N **do**
      $\delta_t(j) = \max_{i \in \{1,\cdots,N\}} [\delta_{t-1}(i) P([S_t = j] \mid [S_{t-1} = i])] P(O_t \mid [S_t = j])$
      $\psi_t(j) = \arg\max_{i \in \{1,\cdots,N\}} [\delta_{t-1}(i) P([S_t = j] \mid [S_{t-1} = i])]$
    end
  end
  $S_T^* = \arg\max_{i \in \{1,\cdots,N\}} [\delta_T(i)]$
  for $t = T - 1$ *down to* 1 **do**
    $S_t^* = \psi_{t+1}(S_{t+1}^*)$
  end
  return $S_{1:T}^*$
end

4.3.2.4 Example

Returning to our example, let us suppose that we have a sequence of temperature readings for five consecutive days $O_{1:5} = \{23.05, 17.5, 18.8, 21.22, 24.32\}$ and that we need to know which sequence of states is more likely to correspond to that sequence. For this, we apply Viterbi's algorithm: first, we apply expressions (4.11) and (4.12) to compute the following table:

t	prior	1	2	3	4	5
O_t		23.05	17.50	18.80	21.22	24.32
$\delta_t(fresh)$	0.5	1.32e-2	7.42e-5	9.63e-6	5.95e-8	1.80e-09
$\delta_t(cold)$	0.5	2.62e-3	3.45e-4	1.50e-6	9.14e-8	1.25e-10
$\psi_t(fresh)$			fresh	cold	cold	cold
$\psi_t(cold)$			fresh	fresh	fresh	fresh

Then, we remark that $S_5 = fresh$ is the argument that maximizes $\delta_5(S_5)$, hence, we assign $S_5^* = fresh$. From there, we backtrack: $S_4^* = \psi_5(fresh) = cold$, and so on down to $t = 1$. At the end, we obtain the following maximum likelihood sequence:

$$S_{1:5}^* = \{fresh, cold, fresh, cold, fresh\}$$

4.3.2.5 Classification

Another useful question consists in choosing between several HMMs, representing different classes of state sequences. This is a common situation, for example, in speech recognition tasks, where every word is encoded as a different HMM.

Classification consists in finding the HMM which is more likely to correspond to a given observation sequence. Often, this is done by applying a maximum likelihood criterion, which means that the decision is taken using the output of eq. (4.7) to select the model. Nevertheless, other classification criteria may also be applied, we will discuss them in §4.6.

4.3.3 Numerical Stability and HMM Scaling

When implementing HMMs in computers, numerical precision limitations may become an issue during the execution of the Viterbi and forward-backward algorithms. The reason is that these algorithms multiply very long sequences of numbers which are smaller than 1; and the small values that result may exceed the machine representation capabilities. This situation may be observed in our example for the Viterbi algorithm (§4.3.2.3), where values become very small even for a very short sequence of observations.

Solutions to this problem are specific to each algorithm:

Viterbi Algorithm. Since the viterbi algorithm operates on the probability maxima for each state, the problem is solved by using log probabilities. Hence, multiplications become additions, which are much easily handled by standard machine floating point representations.

Forward-backward Algorithms. Whereas the forward and backward algorithms perform both multiplications and sums on probabilities, applying logarithms is not a viable solution. Several solutions exist [*eg* Minka, 1999, Mann, 2006], perhaps the most popular is the one described by Rabiner

[1990], which consists in multiplying forward probabilities by *scaling coefficients* in order to keep them in the dynamic range of the machine. These coefficients are computed by normalizing the forward probabilities for every time value:

$$c_t = \frac{1}{\sum_{i=1}^{N} \alpha_t(i)} \tag{4.13}$$

So that the scaled forward probabilities are:

$$\bar{\alpha}_t(i) = c_t \alpha_t(i) \tag{4.14}$$

The same set of coefficients that have been computed for α_t are also used to compute the scaled Backward probabilities:

$$\bar{\beta}_t(i) = c_t \beta_t(i) \tag{4.15}$$

4.4 Parameter Learning

Let us have learning data in the form of K observation sequences $O = \{O^1, \cdots, O^K\}, O^k = \{O_1^k, \cdots, O_{T_k}^k\}$. Supposing that the state is observable, it is straightforward to obtain a maximum likelihood estimate of the model's parameters by counting (cf. §2.2.7):

$$\pi_i = \frac{\sum_{k=1}^{K} C([S_1^k = i])}{\sum_{k=1}^{K} \sum_{n=1}^{N} C([S_1^k = n])} \quad \forall\, i \in \{1, \cdots, N\} \tag{4.16}$$

$$\mu_i = \frac{\sum_{k=1}^{K} \sum_{t=1}^{T_k} C([S_t^k = i])\, O_t^k}{\sum_{k=1}^{K} \sum_{t=1}^{T_k} C([S_t^k = i])} \quad \forall\, i \in \{1, \cdots, N\} \tag{4.17}$$

$$\sigma_i^2 = \frac{\sum_{k=1}^{K} \sum_{t=1}^{T_k} C([S_t^k = i])\, (O_t^k - \mu_i)^2}{\sum_{k=1}^{K} \sum_{t=1}^{T_k} C([S_t^k = i])} \quad \forall\, i \in \{1, \cdots, N\} \tag{4.18}$$

$$a_{i,j} = \frac{\sum_{k=1}^{K} \sum_{t=1}^{T_k - 1} C([S_t^k = i]\, [S_{t+1}^k = j])}{\sum_{k=1}^{K} \sum_{t=1}^{T_k - 1} C([S_t^k = i])} \quad \forall\, i,\ j \in \{1, \cdots, N\} \tag{4.19}$$

Unfortunately, state is hidden, which means that counting may not be used. The standard solution is a specialization of the Expectation-Maximization algorithm [Dempster et al., 1977] known as the Baum-Welch algorithm.

4.4.1 The Baum-Welch Algorithm

This algorithm, originally introduced by Baum et al. [1970] to estimate parameters on a single sequence of discrete observations, was later extended for

Algorithm 4. *Baum_Welch*$(O^{1:K}, \lambda)$

input : A set of observation sequences $O^{1:K}$
Initial estimate of HMM parameters λ
output : HMM parameters $\lambda = \{\pi, b, A\}$

begin
  $converged = false$
  while *not converged* **do**
    Compute forward (α_t^k) and backward (β_t^k) probabilities as well as probability P_{O^k} of every observation sequence O^k
    for $i \in \{1, \cdots, N\}$ **do**

$$\bar{\pi}_i = \frac{\sum_{k=1}^{K} \frac{1}{P_{O^k}} \alpha_1^k(i)\ \beta_1^k(i)}{K}$$

$$\bar{\mu}_i = \frac{\sum_{k=1}^{K} \frac{1}{P_{O^k}} \sum_{t=1}^{T_k} \alpha_t^k(i)\ \beta_t^k(i)\ O_t^k}{\sum_{k=1}^{K} \frac{1}{P_{O^k}} \sum_{t=1}^{T_k} \alpha_t^k(i)\ \beta_t^k(i)}$$

$$\bar{\sigma}_i^2 = \frac{\sum_{k=1}^{K} \frac{1}{P_{O^k}} \sum_{t=1}^{T_k} \alpha_t^k(i)\ \beta_t^k(i)\ (O_t^k - \mu_i)^2}{\sum_{k=1}^{K} \frac{1}{P_{O^k}} \sum_{t=1}^{T_k} \alpha_t^k(i)\ \beta_t^k(i)}$$

      for $j \in \{1, \cdots, N\}$ **do**

$$\bar{a}_{i,j} = \frac{\sum_{k=1}^{K} \frac{1}{P_{O^k}} \sum_{t=2}^{T_k} \alpha_{t-1}^k(i) P([S_t=j]|[S_{t-1}=i]\ \lambda) P(O_t^k|[S_t=j]\ \lambda) \beta_t^k(j)}{\sum_{k=1}^{K} \frac{1}{P_{O^k}} \sum_{t=2}^{T_k} \alpha_{t-1}^k(i) \beta_{t-1}^k(i)}$$

      end
    end
    if $\lambda = \{\bar{\pi}, \bar{A}, \bar{b}, \}$ **then**
      $converged = true$
    end
    else
      $\lambda = \{\bar{\pi}, \bar{A}, \bar{b}, \}$
    end
  end
end

multiple observation sequences by Levinson et al. [1983] and for continuous observations by Juang et al. [1986].

The basic idea of the Baum-Welch algorithm is to estimate $P(S_{1:T} \mid O^k)$ using inference and to use the expected event counts as Expected Sufficient Statistics ESS to obtain a new estimate of the model's parameters. This may be regarded as replacing the count functions that appear in eq. (4.16) to (4.19) by their corresponding probabilities, inferred from the current model parameters λ and observation sequence O^k. For the sake of efficiency, inference is performed using forward-backward probabilities.

Since expected counts and model parameters are interdependent, the computation is iterated. As shown by Baum et al. [1970] the algorithm is guaranteed to converge to a local maximum of data likelihood.

4.4.1.1 Analysis

The algorithm makes extensive use of the forward-backward probabilities in lines 6 through 10:

- Line 6 computes the state prior using the expected counts, which are calculated using (4.9):

$$\bar{\pi}_i = \frac{\sum_{k=1}^{K} P([S_1 = i] \mid O^k\ \lambda)}{\sum_{k=1}^{K} \sum_{S_1} P(S_1 \mid O^k\ \lambda)} = \frac{\sum_{k=1}^{K} \frac{1}{P_{O^k}} \alpha_t^k(i)\ \beta_t^k(i)}{K} \tag{4.20}$$

- Lines 7 and 8 also use the expected counts, obtained with (4.9) to compute the observation probability mean values and covariances:

$$\bar{\mu}_i = \frac{\sum_{k=1}^{K} \sum_{t=1}^{T_k} P([S_t = i] \mid O^k\ \lambda)\ O_t^k}{\sum_{k=1}^{K} \sum_{t=1}^{T_k} P([S_t = i] \mid O^k\ \lambda)} = \frac{\sum_{k=1}^{K} \frac{1}{P_{O^k}} \sum_{t=1}^{T_k} \alpha_t^k(i)\ \beta_t^k(i)\ O_t^k}{\sum_{k=1}^{K} \frac{1}{P_{O^k}} \sum_{t=1}^{T_k} \alpha_t^k(i)\ \beta_t^k(i)} \tag{4.21}$$

$$\begin{aligned} \bar{\sigma}_i^2 &= \frac{\sum_{k=1}^{K} \sum_{t=1}^{T_k} P([S_t = i] \mid O^k\ \lambda)\ (O_t^k - \mu_i)^2}{\sum_{k=1}^{K} \sum_{t=1}^{T_k} P([S_t = i] \mid O^k\ \lambda)} \\ &= \frac{\sum_{k=1}^{K} \frac{1}{P_{O^k}} \sum_{t=1}^{T_k} \alpha_t^k(i)\ \beta_t^k(i)\ (O_t^k - \mu_i)^2}{\sum_{k=1}^{K} \frac{1}{P_{O^k}} \sum_{t=1}^{T_k} \alpha_t^k(i)\ \beta_t^k(i)} \end{aligned} \tag{4.22}$$

- Line 10 computes the expected transition counts using (4.10) an uses it to estimate transition probabilities:

$$\begin{aligned} \bar{a}_{i,j} &= \frac{\sum_{k=1}^{K} \sum_{t=2}^{T_k} P([S_{t-1} = i]\ [S_t = j] \mid O^k \lambda)}{\sum_{k=1}^{K} \sum_{t=2}^{T_k} P([S_{t-1} = i] \mid O^k \lambda)} \\ &= \frac{\sum_{k=1}^{K} \frac{1}{P_{O^k}} \sum_{t=2}^{T_k} \alpha_{t-1}^k(i) P([S_t = j \mid [S_{t-1} = i]\ \lambda) P(O_t^k \mid [S_t = j]\ \lambda) \beta_t^k(j)}{\sum_{k=1}^{K} \frac{1}{P_{O^k}} \sum_{t=2}^{T_k} \alpha_{t-1}^k(i) \beta_{t-1}^k(i)} \end{aligned} \tag{4.23}$$

Since the forward-backward probabilities are computed only once, and then stored in a table, the cost of computing the state prior and observation probabilities if $O(N|O|)$ where $|O|$ is the total number of observations in training data; in a similar way, the cost of computing the transition probability is $O(N^2|O|)$. However, these costs may be considerably reduced by restricting the model's structure, which is the subject of sections 4.5 and 4.6.

4.4.2 Incremental Algorithms

A number of incremental versions of the Baum-Welch algorithm exist in the literature. Neal and Hinton [1998] propose a simple modification, called incremental EM, which, instead of summing over all the observation sequences in the data set in lines 6 through 4.23, processes only one sequence at a time. This is justified by the fact that, since the initial values of the expected counts are assumed to be inaccurate, it is best to update them as soon as possible. The algorithm stores the sum of the expected counts for all the probability distributions, and, in the case of HMMs, it is guaranteed to converge to a local maximum of data likelihood.

Another approach has been proposed by Singer and Warmuth [1996]. Instead of storing expected counts and maximizing the data likelihood $P(O^k_{1:T_k} \mid \lambda^t)$, they maximize an alternative objective function $F(\lambda)$, which penalizes large changes in parameters:

$$F(\lambda^{t+1}) = \lambda P(O^k_{1:T_k} \mid \lambda^t) - d(\lambda^t, \lambda^{t+1}) \tag{4.24}$$

where $d(\lambda^t, \lambda^{t+1})$ is a measure of the difference between the current and the reestimated parameters.

4.5 Transition Structure

Until now, in all our complexity computations for inference and learning algorithms, we have considered a fully connected or *ergodic* HMM (fig. 4.3(a)) where every state may be reached from any other state in a single time step. As we have seen in §4.3.1, this implies the need to iterate through N^2 possible transitions when computing the prediction step. However, for some applications a better representation may be built by assuming that certain transitions are not possible. This is equivalent to constraining the structure of the transition matrix by forcing the value of a subset of the $a_{i,j}$ elements to be zero. This is often called the *structure* or *topology* of the HMM, and, even if they are somewhat related, it should not be confused with the structure of a Bayes network.

We define the structure Φ of an HMM as the specification of the number of states N and the subset of elements in A such that $a_{i,j} = 0$. In other words, Φ specifies the cardinality of the state variable and a subset of forbidden transitions and leaves all other transitions unspecified (but constrained to be strictly positive). A complete model specification consists of the model's structure and parameters $\mathcal{M} = \{\Phi, \lambda\}$.

One of the most popular HMM structures are called left-right models (fig. 4.3(b)). Their fundamental property is that their transitions coefficients obey the following constraint:

$$a_{i,j} = 0 \quad \forall j < i \tag{4.25}$$

It is easier to visualize the HMMs structure as a graph (fig. 4.3), where nodes represent states and for every $a_{i,j} \neq 0$ there is a directed edge from i to j indicating that that transition is allowed. In this case, state j is said to be a *neighbor* of state i.

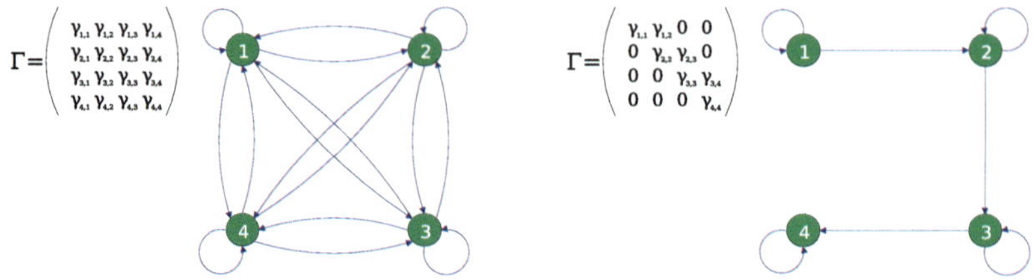

(a) ergodic (b) left-right

Fig. 4.3 Examples of HMM topologies. Both the transition matrix and the corresponding graph are shown.

Of course, other topologies than ergodic or left-right models may be imagined. An interesting case is when A is composed primarily by zeroes (*ie* sparse). In those cases, it is more efficient to represent the transition matrix as an array of linked lists, one list per state (*ie* a transition list). Each linked list stores the neighbors of the corresponding state (fig. 4.4). This representation is not only more compact, it allows to compute the prediction step of filtering in $O(E)$ where E is the number of non-zero elements in the transition matrix.

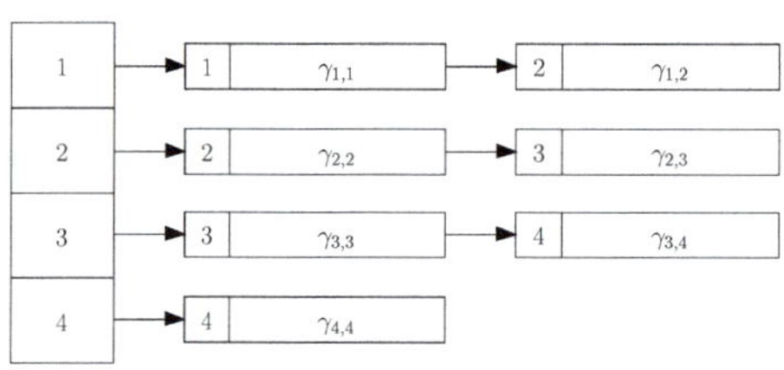

Fig. 4.4 A linked list representation of fig. 4.3(b)

The performance gain may be illustrated by an example. Let us have a 5×5 grid, where every cell represents a state. By only allowing transitions between cells having a common edge [4], we reduce the number of possible transitions from $125^2 = 15625$ for an ergodic model to just 80 for the 4-neighborhood criterion (fig. 4.5).

Defining a non-ergodic topology for the model does not only lead to improved time and memory efficiency, but also influences the quality of inference [cf. Brand, 1998, Freitag and McCallum, 2000, Binsztok and Artières, 2005] and accelerates learning by reducing the number of parameters. But, how to choose the best structure for a given application or problem? It seems natural to do the same as for the models parameters and to learn the structure from data, which is the topic of the following section.

[4] This is called 4-neighborhood, because every cell which is not in the border has exactly four neighbors under this criterion.

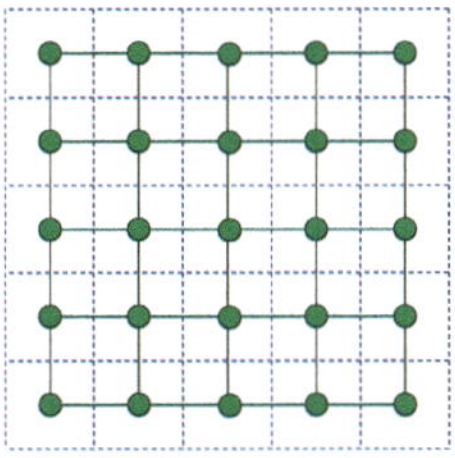

Fig. 4.5 A 5 × 5 grid (dotted blue lines) and the corresponding non directed topology graph (solid green lines) assuming 4-neighborhood.

4.6 Structure Learning

A good starting point to illustrate structure learning is to consider the full Bayesian learning problem, which consists of computing the posterior probability of the model given the data:

$$P(\mathcal{M} \mid O) = \frac{P(O \mid \mathcal{M})P(\mathcal{M})}{P(O)} \tag{4.26}$$

But $P(\mathcal{M} \mid O)$ gives the joint posterior of the structure and the parameters, and we are interested in the posterior probability of the structure alone. Thus, we marginalize over the space of all possible parameters:

$$P(\Phi \mid O) = \frac{P(\Phi)P(O \mid \Phi)}{P(O)} \tag{4.27}$$

$$= \frac{P(\Phi) \int_\lambda P(\lambda \mid \Phi)P(O \mid \Phi\ \lambda)d\lambda}{P(O)} \tag{4.28}$$

Which gives us a probability distribution over all the possible structures. However, computing the full probability distribution is almost never done in practice. Instead, learning is performed by finding a Maximum *a Posteriori* (MAP) estimate of the structure, *ie* the single model that maximizes eq. (4.28).

Although it is tempting to simplify further and apply a Maximum Likelihood criterion by assuming that $P(\mathcal{M})$ is uniform, Maximum Likelihood (ML) estimation suffers from a bias towards complex models which renders it unusable for structure learning[5]: it is trivial to construct a maximum likelihood structure by representing every observation in O with a different discrete state.

Hence, it is necessary to maximize eq. 4.28, which is difficult because no closed form solution for the integral is known [Murphy, 2002]; a common

[5] Maximum likelihood has also been subject to more fundamental critics concerning its semantic and mathematical soundness [Irony and Singpurwalla, 1997].

workaround is to use an approximation of the integral's value known as the Bayes Information Criterion (BIC) [Schwarz, 1978], which is expressed as a log-likelihood:

$$\log P(O \mid \Phi) \approx BIC(O, \Phi, \hat{\lambda}) = \log P(O \mid \Phi\ \hat{\lambda}) - \frac{d}{2} \log|O| \tag{4.29}$$

Where $\hat{\lambda}$ is the maximum likelihood estimate of the parameters given the current structure – obtained, for example, using the Baum-Welch algorithm; d is the number of free parameters in the model; and $|O|$ is the number of observations in training data. The first term is just the data likelihood (4.7) and the second one may be understood as a penalty for model complexity. The BIC converges asymptotically to the log data likelihood as $|O|$ grows.

Taking logarithms in (4.28) and replacing $\log P(O \mid \Phi)$ by the BIC, we obtain:

$$\log P(\Phi \mid O) \approx \log P(\Phi) + BIC(O, \Phi, \hat{\lambda}) - \log P(O) \tag{4.30}$$

Hence, the learning problem becomes that of maximizing (4.30) instead of (4.28). In general, the term $\log P(O)$ is also excluded from the maximization because it does not depend on the structure. A further simplification is to assume a flat prior for the structure $P(\Phi)$, given that complexity is already penalized by the BIC. Thus, in many approaches, just the BIC is maximized.

Maybe the most important difficulty in structure learning lies in the huge size of the space of structures. Even for a fixed number of states N, the number of possible adjacency matrices is $O(2^{N^2})$. This makes it necessary to restrict the structure to a smaller family of graphs (*eg* trees, directed acyclic graphs) [*eg* Pearl, 1988, Meila et al., 2001]; an alternative is to conduct an heuristic based search in structure space.

4.6.1 Local Search Algorithms

Due to the complexity of structure estimation, instead of using a global algorithm to maximize (4.30), a local search is often conducted, where the algorithm starts with an initial guess of Φ and then searches in its *neighborhood* for structures having a higher BIC. Neighbors are obtained by applying simple operators to Φ such as adding, deleting or reversing an edge (fig. 4.6). Depending on the algorithm, one or more of those structures are selected and the process is iterated.

Despite the use of approximations like the BIC, local search strategies are still expensive, mainly because the EM (Baum-Welch) algorithm must be run for every candidate structure. This has motivated a different approach proposed by Friedman [1997]: instead of running the EM algorithm inside the search procedure, the search is performed inside EM (alg. 5). This is known as *Structural EM*.

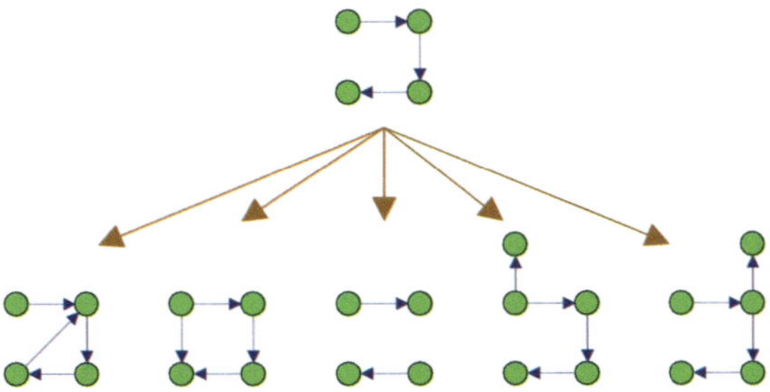

Fig. 4.6 Local structure search: neighbors are found by adding or removing nodes and edges.

Algorithm 5. *Structural_EM*$(O^{1:K}, \lambda, \Phi)$

---

input : A set of observation sequences $O^{1:K}$
Initial estimate of HMM parameters λ
Initial structure estimate Φ

output : HMM parameters $\lambda = \{\pi, b, A\}$
HMM Structure Φ

1 **begin**
2   *converged* = *false*
3   **while** *not converged* **do**
4   Improve λ using EM
5   **for** *each neighbor* Φ' *of* Φ **do**
6   Compute expected counts for Φ' using Φ and λ ; `/* Structural E-Step */`
7   Compute $\hat{\lambda}'$ using the expected counts
8   Compute $BIC(O^{1:K}, \Phi', \hat{\lambda}')$
9   **end**
10         $\Phi^* = \arg\max_{\Phi'} BIC(O^{1:K}, \Phi')$
11   **if** $BIC(O^{1:K}, \Phi^*, \hat{\lambda}') > BIC(O^{1:K}, \Phi, \lambda)$ **then**
12             $\Phi = \Phi'$; `/* Structural M-Step */`
13             $\lambda = \hat{\lambda}'$
14   **end**
15   **else**
16   *converged* = *true*
17   **end**
18   **end**
19 **end**

The key to structural EM lies in steps 7 and 8. The estimation of $\hat{\lambda}'$ is performed by updating $\hat{\lambda}$ according to the difference between Φ and Φ', which, as we mentioned above, consists of a small change such as and added or deleted edge. Hence, it is not necessary to run the whole Baum-Welch algorithm for every neighbor, but only for the one that gets the best BIC score in every iteration. The algorithm is guaranteed to converge to a local maximum of the BIC score.

4.6.2 State Merging Algorithms

State merging algorithms [Stolcke and Omohundro, 1994, Seymore et al., 1999, Binsztok and Artières, 2005] work by applying heuristics which are very similar to agglomerative clustering. First, they assume that every observation in O corresponds to a different state, and assigns a probability of one to edges connecting states corresponding to consecutive observations and zero to all others. The result of this step is a maximum likelihood model, but which is overly complex and do not generalize at all.

From there, the algorithm builds a simpler– and more general – model by merging states (see fig. 4.7). At every iteration, the algorithm searches for the merging which gives the highest increase on the posterior (4.28) and it stops when no further increase is detected.

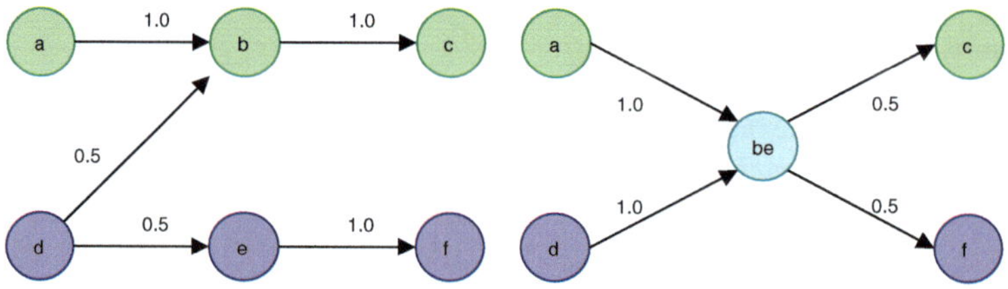

(a) before (b) after

Fig. 4.7 Merging states b and e produces state be. Notice how the probabilities associated to incoming edges are summed, while those associated to outgoing edges are renormalized.

In order to reestimate the model parameters, model merging uses a simplified version of the Baum-Welch algorithm, which only reestimates the parameters for the most probable path – which is computed using the Viterbi algorithm – instead than for the whole model; thus reducing computation time at the expense of optimality. However, experimental results seem to be comparable to those obtained using Baum-Welch.

Model merging is able to estimate the number of states N in an intuitive fashion. Moreover, it seems possible to adapt it to work incrementally. On the downside, model merging is very sensitive to differences in temporal alignment and it has been mostly tested on applications having discrete observations.

4.6.3 Other Algorithms

Another interesting approach has been proposed by Brand [1998], instead of searching or building the structure, he transforms the Baum-Welch algorithm into a MAP estimation procedure called Entropic-EM by incorporating an entropic prior in the computation of the new parameters. This prior favors informative values (those that are near from zero or one) when training data is scant, but converges to the conventional maximum likelihood estimation estimate as more observations are available. The result of using this MAP criterion is that irrelevant parameters are driven asymptotically to zero. After Entropic-EM learning, edges or states which are considered not informative are trimmed from the structure.

A somewhat similar idea has been proposed by Vasko et al. [1997]. They first train a fully connected HMM using Baum-Welch. From there, they iteratively remove transitions or states from the model. Since the fully connected model has maximum data likelihood, every iteration decreases it. The chosen structure is the simplest one before a substantial decrease in the likelihood.

The idea of starting with a very simple model to which new nodes and transitions are added during learning – hence, incrementing data likelihood – has also been explored by a number of authors Lockwood and Blanchet [1993], Freitag and McCallum [2000], but these approaches are not general, since they all apply domain-specific knowledge.

4.7 Discussion

In this chapter we explained several basic concepts related to Hidden Markov Models. We presented the probabilistic model: variables, JPD and parametric forms, making emphasis in continuous observation HMMs. We also presented the “classic” inference and learning algorithms for HMMs, showing that maintaining the belief state constitutes the basis for inference. Correspondingly, we illustrated the use of dynamic programming to perform efficient off-line inference. We outlined the use of the Baum-Welch algorithm, which, basically consists in using off-line learning to compute the ESS which are necessary to estimate the model's parameters. Finally, we have introduced the problem of structure learning, and presented an overview of the existing approaches.

We would like to conclude with some reflections about the use of HMMs in the context of pattern based motion prediction. As we have seen in chapter 3, HMMs are very popular in the literature of pattern based motion prediction, but, what are the reasons of this popularity? indeed, there are many reasons: First, HMMs are a discrete model, which makes them simpler to manipulate than a continuous representation. Moreover, the existence of efficient and well tested algorithms for inference and learning makes their application straightforward.

On the other hand, in spite of the efficiency of existing algorithms, the computational cost of inference on conventional ergodic Hidden Markov Models grows with the square of the number of discrete states in the model; which makes it too complex to scale their use to problems which require hundreds or even thousands of states and huge data sets.

There exist at least three ways to address this problem: a) using hierarchical representations such as Abstract Hidden Markov Models Bui et al. [2002] which decompose the problem using a divide and conquer strategy; b) defining simpler transition structures which, at the same time reduce the algorithmic complexity and provide a better representation of motion; and c) using approximate inference tools like particle filters [Arulampalam et al., 2002].

But these techniques do not solve the problems of parameter and, in particular, structure learning which is hard even in the off-line case. At the same time, approximate inference tools are not necessary provided that learned structures are simple enough to allow real-time inference.

Recapitulating, for a working lifelong learning approach based on HMMs, we want a learning algorithm that satisfies the following conditions:

1. Both parameters and structure should be learned.
2. Learning should be incremental.
3. The learned structure should be both meaningful and simple enough to allow inference in real-time.

Despite the existence of incremental extensions of the Baum-Welch and of structure learning algorithms, for the best of our knowledge there exists only one algorithm – state merging – which is able to simultaneously learn the parameters *and* the structure of an HMM in an incremental way. Unfortunately, they are defined only for discrete observations[6] and, in addition, there is no guarantee that the structures obtained by applying model merging will be simple and semantically sound. Therefore, in order to fulfill the third condition, we have opted for designing a new learning algorithm, which is the subject of chapter 5.

[6] With the possible exception of the recent work of Kulic et al. [2008], which apply an algorithm similar to model merging to continuous multidimensional observations.

Part III
Proposed Approach

Chapter 5
Growing Hidden Markov Models

It's all to do with the training: you can do a lot if you're properly trained

QUEEN ELIZABETH II
Television documentary

5.1 Overview

As we have shown in the first part of this book, Hidden Markov Models constitute a powerful probabilistic tool. Unfortunately, the existing learning algorithms for HMMs are not well suited for lifelong learning. This chapter introduces a tool that extends HMMs to address this weakness and which constitutes our main contribution: Growing Hidden Markov Models (GHMMs). They can be described as time-evolving HMMs with continuous observation variables where the number of states, the structure and the probability parameters are updated every time that an observation sequence is available. The learning algorithm verifies the three properties that, according to Langley [1995], define an incremental learning algorithm: a) it operates on one observation sequence at a time, b) it does not reprocess any previous data, and c) it retains only one knowledge structure in memory.

Our approach is designed for its utilization as a discrete approximate inference tool for continuous state spaces. Although we use it here in the context of motion prediction, it is designed to be general and can be applied to any problem where the following assumptions are true:

- The input has been priorly segmented into observation sequences corresponding to complete instances of the process or system being modeled (*eg* in our application the input is segmented into pedestrian trajectories).
- The evolution of the state of the modeled system or process is a continuous function.

A.D. Vasquez: Incre. Learn. for Motion Predict. of Pedest. & Vehicles, STAR 64, pp. 71–82.
springerlink.com

- The observation space is a subspace of the continuous state space. This implies that, by finding a decomposition of the observation space, a decomposition is also performed on the continuous state space[1]

The key intuition behind GHMMs is that the structure of the model should reflect the spatial structure of the state space discretization, where transitions between states are only allowed if the corresponding regions are neighbors. Therefore, structure learning basically consists of estimating the best space discretization from data and identifying neighboring regions. We have addressed this problem by building a *topological map* using the Instantaneous Topological Map (ITM) algorithm [Jockusch and Ritter, 1999]. For parameter learning, we have adapted the incremental Expectation-Maximization approach proposed by Neal and Hinton [1998] in order to deal with a changing number of discrete states and with continuous observations.

To avoid confusion, in the rest of this book, we make a strict distinction between: the *nodes* of the Instantaneous Topological Map (ITM) algorithm, the *discrete states* of a GHMM, the *continuous state* of an object and the *observations* provided by sensors.

The following section presents GHMMs as a probabilistic model, and discusses its use for inference. Then, we outline the proposed learning algorithm, introduce the topological map, and detail the algorithm's steps. We conclude by enumerating the properties of the learning algorithm and describe some possible ways of improving it.

5.2 Probabilistic Model

Structurally, GHMMs are identical to regular HMMs except for the fact that the number of states and the transition structure are not constant, but change as more input observation sequences are processed.

5.2.1 Variables

As for HMMs, a GHMM is defined in terms of three variables:

- S_t, S_{t-1}, the current and previous states, which are discrete variables with value $S_t, S_{t-1} \in \{1, \cdots, N_k\}$, where N_k is the number of states in the model after k observation sequences have been processed.
- O_t, the observation variable, which is a multidimensional vector.

[1] It is worth noting that this assumption may be relaxed when the model is not used for prediction but – for instance – just for recognition. In that case the only requirement is the existence of a weak topological equivalence or topographic mapping between the observation and state spaces; when the system goes through states which are near each other, the corresponding observations will also be close to each other.

The sequence index k highlights the fact that the parameters and structure of the model change with every new observation sequence. Nevertheless, since GHMMs operate on a single observation sequence at a time, this index becomes redundant in the notation, and we will omit it hereafter.

5.2.2 Decomposition

The joint probability decomposition (JPD) for GHMMs is the same we have already discussed for the Bayes filter:

$$P(S_{t-1}\ S_t\ O_t) = P(S_{t-1})P(S_t \mid S_{t-1})P(O_t \mid S_t) \tag{5.1}$$

5.2.3 Parametric Forms

The parametric forms here are essentially the same than for regular HMMs, except for the fact that we restrict all the Gaussians of the observation probabilities to have the same covariance matrix.

- $P(S_0 = i) = \pi_i$. The state prior will be represented as a vector $\pi = \{\pi_1, \cdots, \pi_N\}$ where each element contains the prior probability for the corresponding discrete state.
- $P([S_t = j] | [S_{t-1} = i]) = a_{i,j}$. Transition probabilities are represented with a set of variables A, where each element $a_{i,j}$ represents the probability of reaching state j in the next time step given that the system is currently in state i. Matrices are not used anymore because the learned structure is sparse, justifying the use of alternative structures like the linked lists arrays we have introduced in Sec. 4.5.
- $P(O_t | [S_t = i]) = \mathbf{G}(O_t;\ \mu_i,\ \Sigma)$. The observation probability density function will be represented by a Gaussian distribution for every discrete state, having the same covariance matrix Σ for all discrete states. The set of all the Gaussians' parameters will be denoted as $b = \{\Sigma, \mu_1, \cdots, \mu_N\}$.

5.3 Inference

Since GHMMs are HMMs, they may answer the same probabilistic questions (see §4.3). In this section we will review the two basic questions that should be answered in the context of our application: filtering and prediction.

The goal of filtering is is to update the belief state of an object on the basis of an observation. This is done using expression (4.3) which we copy here for reference:

$$P(S_t \mid O_{1:t}) = \frac{1}{Z} P(O_t \mid S_t) \sum_{S_{t-1}} [P(S_t \mid S_{t-1}) P(S_{t-1} \mid O_{1:t-1})]$$

where $P(S_{t-1} \mid O_{t-1})$ is the belief state calculated in the previous time step. Once the belief state has been estimated, the predicted state H time steps ahead from the present is computed with:

$$P(S_{t+H} \mid O_t) = \sum_{S_{t+H-1}} P(S_{t+H} \mid S_{t+H-1}) P(S_{t+H-1} \mid O_t) \tag{5.2}$$

5.4 Learning

The learning algorithm is what really differentiates GHMMs from HMMs. It processes one observation sequence at a time and consists of three steps (Fig. 5.1):

1. *Topological Map Update.* For every observation in a sequence, our approach uses the Instantaneous Topological Map to update a graph describing the topology of the observation space. The update may add or delete nodes and edges, and modify the reference vectors associated to the ITM nodes.
2. *GHMM Structure Update.* Changes in the topological map are reflected in the GHMM structure. New transition and prior probabilities are initialized using preset values.
3. *Parameter Update.* Using the whole observation sequence and the updated GHMM structure, a single iteration of the incremental Baum-Welch algorithm is run in order to update the model's parameters.

In the rest of this section we introduce the topological map, and provide the details of the three steps of the learning algorithm, finally, we discuss the properties of the algorithm and and the issues involved in learning the covariance matrix.

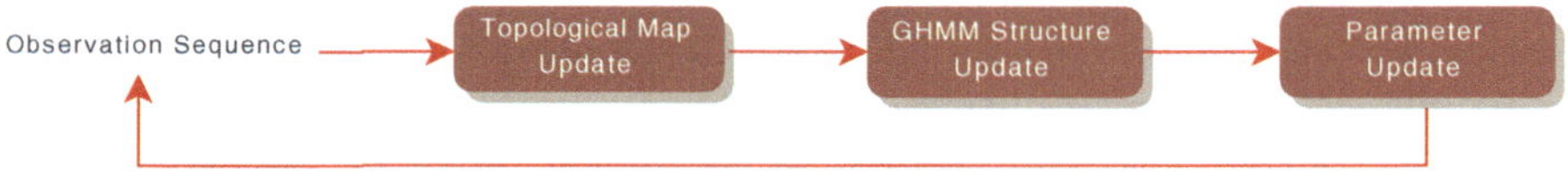

Fig. 5.1 Overview of the GHMM learning algorithm.

5.4.1 Topological Map

In the context of this book, what we call a topological map is a discrete representation of the state-space in the form of a graph, where nodes represent discrete regions and edges are used to indicate explicitly which regions are contiguous[2].

[2] It is important to note that, for us, a topological map is essentially a metric representation in which neighborhood is represented explicitly. This differs from the concept of a topological map we have discussed in § 3.6.2, which have a coarser representation often based on higher level semantics [*eg* Kuipers, 1998].

We will illustrate the concept with a simple grid discretization of a two dimensional space (Fig. 5.2(a)). The corresponding topological map will have one node for every cell in the grid, and edges between cells that share a border, for the sake of simplicity, we do not consider diagonal cells to be neighbors.

The state is assumed to evolve as a continuous function of the time. Thus, if we plot an object's motion, it should inevitably pass through cells borders as it goes from one discrete region to another (Fig. 5.2(b)). Since, by definition, there is one edge for every border, it follows that the best approximation (Fig.5.2(c)) to any possible continuous trajectory may be built using only edges that belong to the graph.

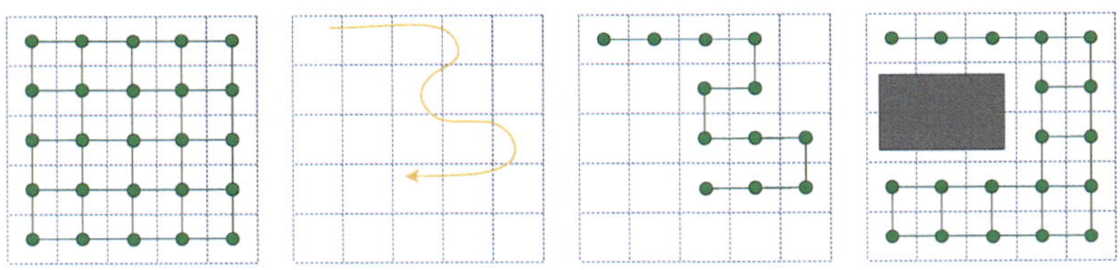

(a) grid and graph (b) trajectory in space (c) trajectory in the graph (d) an obstacle

Fig. 5.2 a) Example of a discretization of space into a grid of rectangular cells (blue dotted lines), and the corresponding topological map representation (green dots and solid lines); b) a continuous trajectory; c) the same trajectory represented as a succession of nodes and edges in the map; and d) an obstacle (gray rectangle) and the corresponding topological map.

Often, the process evolves in a small manifold of the state space. For example, moving objects tend to move frequently in some regions (*eg* corridors) while never going into others (*eg* obstacles). This can not be captured by static decompositions such as grids because they are chosen in advance. Therefore, the whole space is represented, with the consequent waste of computational resources (see § 3.4). It would be preferable to adapt the discretization to the observed data in order to approximate the actual manifold in which state evolves, so that unused regions of the space are not represented (see Fig. 5.2(d)). This poses the problem of how to perform this discretization and, at the same time, identify neighbor regions in an efficient and incremental way.

Our solution to this problem uses the Instantaneous Topological Map, an efficient Vector Quantization algorithm, which is able to learn the spatial distribution of data along with its topology. The application of the algorithm is explained in the following section, the interested reader may consult Appendix A for a general introduction to Vector Quantization and Topology Representing Networks.

5.4.2 Updating the Topological Map

The topological map is represented as a graph where every node i has an associated vector w_i, corresponding to the region's centroid. The nodes are added an adapted in order to minimize the distortion of the model, *ie* the sum of the squared distances between the input (*ie* observation) vectors and the centroid of their closest node.

The topological map is updated for every available observation O_t using the ITM algorithm which has the following properties:

- It minimizes the number of nodes while trying to keep always a uniform distance between neighbors.
- Has linear time and memory complexity with respect to the number of nodes.
- Edges are a subset of the Delaunay triangulation, which implies that they can exist only between nodes representing adjacent Voronoi[3] regions (Fig. 5.3).

The ITM algorithm consists of the following steps (*cf* [Jockusch and Ritter, 1999]):

1. **Matching:** find the nearest b and second nearest s nodes to O_t. We use the Mahalanobis distance – Eq. (5.3) – with the same Σ than for observation probabilities.

$$d^2_\Sigma(u,v) = (u-v)^T \Sigma^{-1}(u-v) \tag{5.3}$$

   where u and v are two reference vectors.
2. **Weight adaptation:** move w_b towards O_t by a small fraction $\Delta_b = \epsilon(O_t - w_b)$.
3. **Edge adaptation:** a) create an edge connecting b and s unless that edge exists; b) for every neighbor m of b check if O_t lies in the Thales sphere going through w_m and w_b and delete the corresponding edge if that is the case. Delete any node which has no neighbors.
4. **Node adaptation:** a) if O_t lies outside the Thales sphere going through w_b and w_s and its distance from w_b is greater than a given threshold τ, create a new node n with $w_n = O_t$. Connect nodes b and n. Remove node s if it's distance from b is smaller than $\frac{\tau}{2}$.

A crucial factor for the efficiency of the algorithm is that, besides the matching step, all the operations needed to maintain the Delaunay triangulation depend only on nodes and edges in a local neighborhood. There is a

[3] The Voronoi region associated with a node is defined by the set of all the points which are closer to that node's centroid than to any other centroid in the graph. Delaunay edges link the centroids of Voronoi regions that have a common border(*cf* Appendix A).

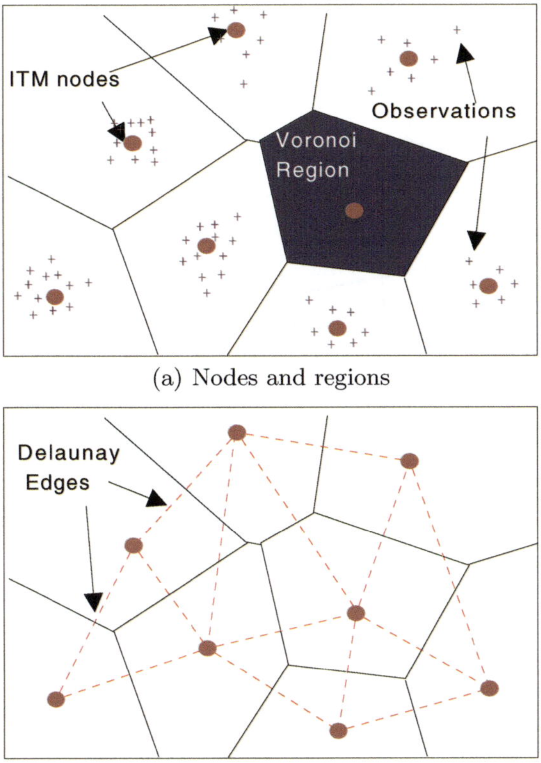

(a) Nodes and regions

(b) Edges

Fig. 5.3 Example ITM space decomposition

minor problem though, since node adaptation takes place after edge adaptation, it is possible that some of the edges connected to b become non Delaunay. However, these edges are later deleted by the edge adaptation step when new observations fall in their neighborhood.

It is important to note that, due to the assumption that the observation space is actually a subspace of the continuous state space, the obtained ITM is also a representation of the latter. This makes it possible to use it directly to update the GHMM structure, as described in the following section.

5.4.3 Updating the Model's Structure

During the topological map update, nodes and edges may be added or deleted, these changes in the topological map are reflected in the GHMM structure as follows:

1. For every new node i in the topological map, add a corresponding discrete state in the GHMM, initializing its prior to a preset value: $\pi_i = \pi_0$. Do the

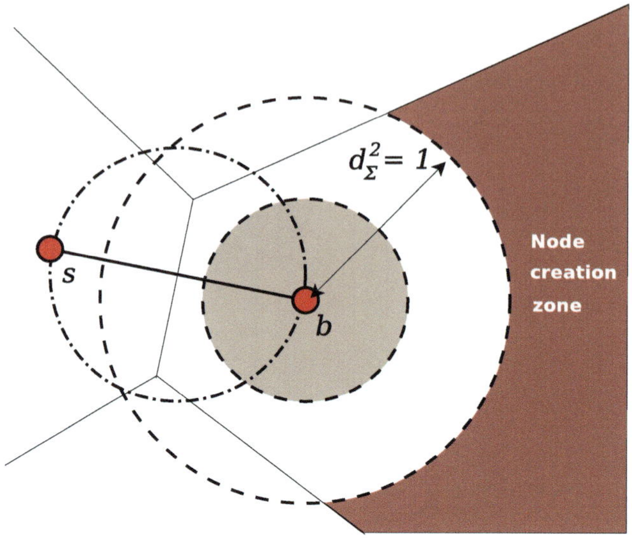

Fig. 5.4 ITM Node adaptation. A node is created when it falls in the light blue zone, this may be followed by the deletion of the second best node if it inside the light green zone.

same for the self-transition probability: $a_{i,i} = a_0$. Note that in this and the two following steps, the values are not strictly a probability because the corresponding sums do not add to one. This is corrected by a normalization step that takes place at the beginning of parameter update (*cf* § 5.4.4).

2. For every new edge (i, j) in the topological map, initialize the corresponding transition weights to: $a_{i,j} = a_0$ and $a_{j,i} = a_0$. As in the previous step, this values will be normalized later to obtain true probabilities.
3. For every deleted node and edge in the topological map, assign a value of zero (*ie* delete) to the corresponding state prior and transition weights.
4. For every added or modified centroid w_i, set the corresponding Gaussian mean value: $\mu_i = w_i$.

5.4.4 Updating the Parameters

Parameter learning takes place once per input sequence, after all the observations have been processed by the structure learning step. The GHMM learning algorithm reestimates the parameters using an incremental version of the Baum-Welch technique based on the work from Neal and Hinton [1998] extending it for continuous observation variables and an evolving number of states. The basic idea of these algorithms is to use inference to compute, for every state an transition, the likelihood that it belongs to the state (or transition) sequence that best explains the observation sequence. Then, these likelihoods are used as weights to update the model.

A particularity of our approach is that all of the observation probabilities' mean values have been assigned during structure update (see § 5.4.3) and that their covariance Σ is fixed. Hence, only the state prior and transition probabilities need to be reestimated. This is done in four steps:

1. Normalize the state prior and transition values. This is necessary because the structure update does not guarantee that the corresponding probabilities add up to one, as explained in § 5.4.3.
2. Precompute α_i (forward probabilities), β_i (backward probabilities) and P_O (joint observation probability) for the observation sequence $O_{1:T}$ (see section 4.3.2).
3. For every discrete state i in the GHMM, reestimate the state prior:
$$\hat{\pi}_i \leftarrow \frac{\alpha_1(i)\ \beta_1(i)}{P_O} \tag{5.4}$$
$$\pi_i \leftarrow \frac{(k-1)\pi_i + \hat{\pi}_i}{k} \tag{5.5}$$
where k is the number of observation sequences that have been observed so far.
4. Reestimate every non-zero transition probability in the GHMM using equations (5.6) and (5.7).
$$\hat{a}_{i,j} \leftarrow \frac{\sum_{t=2}^{T} \alpha_{t-1}(i) a_{i,j} P(O_t | [S_t = j]) \beta_t(j)}{\sum_{t=2}^{T} \alpha_{t-1}(i) \beta_{t-1}(i)} \tag{5.6}$$
$$a_{i,j} \leftarrow \frac{(k-1)a_{i,j} + \hat{a}_{i,j}}{k} \tag{5.7}$$

These steps constitute a single iteration of incremental Baum-Welch. The reason to use Equations (5.5) and (5.7) is that they are equivalent to dividing the sum of the weight by the number of trajectories in order to obtain an average weight value; these equations, together with the preset values π_0 and a_0, are the elements that enable incremental learning with an evolving number of states.

For the sake of comparison, we also performed early tests with straightforward Baum-Welch, *ie* using only (5.4) and (5.6) to make the update, but the learned parameters were too heavily biased towards recent observation sequences.

5.4.5 Properties

5.4.5.1 Convergence

One theoretical drawback of the ITM algorithm is that strict convergence to a local minimum of the distortion is not verified. The reason is that the algorithm, like other Topology Representing Networks, uses a constant learning rate, which leads to models which, instead of converging, keep oscillating around local minima.

However, in the case of ITM it is possible to establish a theoretical upper limit to the distortion, based on the fact that the algorithm guarantees that the maximum Mahalanobis distance to an observation will be equal or inferior to τ.

The distortion $\hat{E}$ may be computed as the average distance from an observation to the reference vector of its closest node (*cf* appendix A):

$$\hat{E} = \frac{1}{|O|} \sum_{i=1}^{K} \sum_{O_t \in \mathcal{V}_i} d^2_{\Sigma}(O_t, w_i) \tag{5.8}$$

where $|O|$ is the number of input vectors or observations in training data. But, since every observation belongs to only one Voronoi region and $d^2_{\Sigma}(O_t, w_i) \leq \tau$, we know that:

$$\hat{E} \leq \tau \tag{5.9}$$

5.4.5.2 Number of Edges

As we have stated throughout this text, we want the learned structure – thus, the topological map – to be both simple and meaningful. We have seen that the topological map is a subset of the Delaunay triangulation, which has a simple interpretation in terms of neighborhood, and seems plausible when partitioning a continuous space into discrete regions.

But, what about simplicity? it turns out that the upper bound on the number of edges is the same than for the Delaunay triangulation, which, for reference vectors in two dimensional space is $O(N)$ because the triangulation's edges and vertices make up a planar graph [Aurenhammer, 1991]. However, for higher-dimensional spaces things seem less promising, for example Dewdney and Vranch [1977] have proved that the upper bound for the size of a Delaunay triangulation in $\mathbb{R}^3$ is $O(N^2)$, which is equivalent to a fully connected graph.

On the other hand, Dwyer [1989] has demonstrated that the expected size of the Delaunay triangulation in any d-dimensional space is $O(N)$, supposing that the reference vectors are drawn uniformly from the unit sphere. As observed by Aurenhammer and Klein [2000], this result indicates that high-dimensional Delaunay triangulations will be small ($O(N)$) in most practical situations.

5.4.5.3 Complexity

The ITM algorithm is designed from the ground up for incremental learning. It processes observations in a one by one basis. The time complexity of the topological map update is $O(N)$. Moreover, since the most expensive operation is the matching step, it seems feasible to improve the efficiency of the algorithm by using a hierarchical space indexing techniques, like R-trees and their extensions [*eg* Guttman, 1984, Beckmann et al., 1990].

5.4.6 Learning the Covariance

One possible extension of the learning algorithm would be to learn the individual covariance matrices of the observation probabilities. As a matter of fact, efficient incremental algorithms exist to perform this computation [*eg* Li et al., 2003, Igel et al., 2006]. However, since the covariances are used to compute the Mahalanobis distance, using theses algorithms would modify the ITM algorithm, with the following consequences:

- Since the distance criterion would not be uniform for all the space, it is not sure that the properties of the ITM algorithm will still hold. Hence, further mathematical analysis would be needed to ensure it.
- Due to the fact that the observations that have been already associated to a node in the ITM determine its covariance matrix and, thus, the dimensions of its Voronoi regions, it is possible to reach a situation where the covariance matrix becomes progressively smaller as time passes.

Some measures may be taken to tackle this problems, for example, it is possible to use a fixed covariance matrix for the ITM algorithm and to learn different covariances for the HMM. However, research in this direction belongs to future work and is out of the scope of this document.

5.5 Discussion

This chapter presented a novel extension to HMMs: Growing Hidden Markov Models, which mainly differs from the standard technique in the fact that the model structure is not fixed, but evolves continuously as more observations are available. Our approach is applicable to those cases in which a Hidden Markov Model is used as a discrete approximation to model a continuous state process.

The main insight behind GHMMs is that, since discrete states are obtained by discretizing the state-space in regions, transitions should be allowed only between neighbor regions. Hence, the problem of structure learning becomes that of building a topological map, by discretizing the continuous state space into discrete regions and identifying neighboring regions. In our approach, this is done by applying the Instantaneous Topological Map, a Topology Representing Network which is well suited for correlated data such as trajectories. By integrating the ITM algorithm with incremental parameter learning, we have been able to build a learning approach which fulfills the three requirements that we have defined at the end of chapter 4:

1. The algorithm learns both the parameters and the structure of the HMM.
2. The algorithm is incremental.
3. The learned structure has an intuitive meaning, and it is simple enough to allow exact inference in real time.

Having presented GHMMs, we are now ready to explain how they may be used in the context of motion prediction, which is the subject of the following chapter.

Chapter 6
Learning and Predicting Motion with GHMMs

– Would you tell me, please, which way I ought to go from here?
– That depends a good deal on where you want to get to – said the Cat.
– I don't much care where– said Alice.
– Then it doesn't matter which way you go – said the Cat.

LEWIS CARROLL
Alice's adventures in wonderland

6.1 Overview

In chapter 5, we have presented GHMMs from a general perspective. In contrast, this chapter focuses on the application of GHMMs as a tool which we use to predict the motion of pedestrians and vehicles. Therefore, the issues addressed here – as well as their respective solutions – are specific to this application and they probably may not be generalized to other domains.

Our application is based on the assumption that people and vehicles move in function of their intention to reach a particular state (*ie* its goal): a car moves in a car park in order to stop at a parking place, a person walks in an office with the intention to reach his desk, etc. Accordingly, we model the object's motion as a sequence of *augmented state vectors*, composed of two sets of variables describing its *current* and *goal* state, respectively.

An important difference between our approach and other HMM based techniques is that motion patterns are not defined in terms of typical trajectories. Instead, they are determined by the object's goal: two objects are involved in the same motion pattern if they intend to reach the same state. In our approach, a motion pattern is no longer a trajectory prototype, but a directed graph indicating all the possible ways in which a goal may be reached[1] as illustrated by Fig. 6.1.

[1] Indeed, our representation bears some resemblance to Markov Decision Processes [Howard, 1960, Cassandra et al., 1996], a popular probabilistic planning tool.

A.D. Vasquez: Incre. Learn. for Motion Predict. of Pedest. & Vehicles, STAR 64, pp. 83–94.
springerlink.com

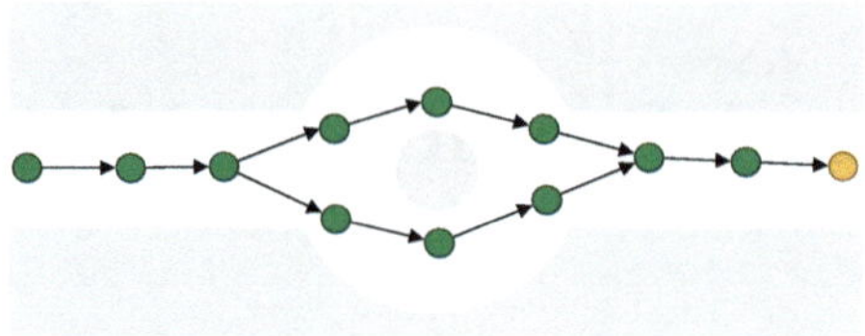

Fig. 6.1 Example of a motion pattern defined in terms of a goal (orange node)

This chapter is organized as follows: in section 6.2, we introduce additional notation and basic assumptions. Section 6.3 explains how the extended state integrates with GHMMs to represent motion patterns. Section 6.4 discusses the use of our model to predict motion, and how, thanks to the extended state, it is possible to predict the goal that an object intends to reach as a side effect of updating the belief state. In section 6.5 we describe the particularities of learning in the context of this application. In section 6.6 we illustrate our approach using an example. Then, in section 6.7, we compare our approach against existing HMM based approaches. Lastly, we present our concluding remarks for this chapter in section 6.8.

6.2 Notation and Basic Assumptions

We assume that tracking data are available as a collection of observation sequences (*ie* trajectories). Every individual sequence $O^k_{1:T^k} = \{O_1, \cdots, O_{T^k}\}$ corresponds to the tracker's output for a single object and its observations are evenly spaced in time. Different observation sequences may have different lengths T^k.

Unless stated otherwise, in the rest of this chapter, we assume that the state of an object is defined by its position and velocities (x, y, x', y') and thus, that the augmented state of the object consists of its current position and velocities, as well as its goal position $(x, y, x', y', \hat{x}, \hat{y})$. It should be noted, however, that our approach is applicable to spaces of arbitrary dimensions.

We assume that observations are available in the form of estimates of the object's coordinates and velocities $O_t = (x_t, y_t, x'_t, y'_t)$ – although, as in the case of the continuous state, is it also possible to include other variables, such as the size or orientation of the object. Since learning is performed on the basis of complete observation sequences, we assume that the position of the last observation $O_T = (x_T, y_T, x'_T, y'_T)$ of each sequence corresponds to the object's goal. Hence, it is possible to build an *augmented observation sequence*, which constitutes the actual input to our algorithm:

$$\bar{O}_{1:T} = \{(x_1, y_1, x'_1, y'_1, x_T, y_T), (x_2, y_2, x'_2, y'_2, x_T, y_T), \cdots, (x_T, y_T, x'_T, y'_T, x_T, y_T)\}$$

6.3 Probabilistic Model

Strictly speaking, our proposed motion model is a standard GHMM which may be manipulated using the techniques that we have already described in 5. However, our definition of an extended state in terms of the current object's state and its goal[2] has implications that need to be analyzed in detail.

Until now, we have assumed that observations coming from sensors are available at every time step. In this application, this assumption is not always true: while it still holds for the current object state, it is no longer true in the case of goals.

By definition, the goal that an object aims to reach may not be observed – either directly or indirectly – until the object has reached it, that is, when its trajectory has ended. On the other hand, this situation does not arise in the case of learning because it takes complete trajectories as input, meaning that the last observation for every trajectory is known and, therefore, that goal observations (*ie* the last observation of a sequence) are available for the learning algorithm.

This section describes how GHMMs are used to build motion models using our extended state definition, it also details how to cope with goal observations, which, as we have shown, are only available at the end of an observation sequence.

6.3.1 Variables

Our approach uses a GHMM, hence, it has the same probabilistic model that has been described in Chapter 5. Nevertheless, we also need to distinguish between the current and intended components of the observation vector. Thus, we decompose the augmented observation variable into a current component – denoted as O'_t – and an intended, or goal component – denoted as O''_t. This results in the following variables:

- S_t, S_{t-1}, defining the current and previous *augmented* states, respectively. These variables take values in $\mathcal{S}_k$, which, as described previously, is the set of model's states after k trajectories have been processed.
- O_t, which describes the current observation. Observations are assumed to be continuous vectors. As explained above, observations are decomposed in two components representing the current and the intended state: $O_t = [O'_t, O''_t]$.

6.3.2 Decomposition

The JPD decomposition is the same as for GHMMs:

---

[2] Hereafter, we will use the terms "intended state" and "goal" interchangeably.

$$P(S_{t-1}\ S_t\ O_t) = P(S_{t-1})P(S_t \mid S_{t-1})P(O_t \mid S_t) \tag{6.1}$$

However, observation variables actually represent the joint occurrence of their current and intended components, which implies that observation probability may be rewritten in terms of these components.

$$P(O_t \mid S_t) = P(O'_t\ O''_t \mid S_t) \tag{6.2}$$

we will assume that the current and intended components of the observations are conditionally independent given the current state:

$$P(O'_t\ O''_t \mid S_t) = P(O'_t \mid S_t)P(O''_t \mid S_t) \tag{6.3}$$

which permits us to rewrite the JPD in terms of the components of the observation variable:

$$P(S_{t-1}\ S_t\ O'_t\ O''_t) = P(S_{t-1})P(S_t \mid S_{t-1})P(O'_t \mid S_t)P(O''_t \mid S_t) \tag{6.4}$$

6.3.3 Parametric Forms

The parametric forms are essentially the same that for GHMMs (see chapter 5), with some added restrictions for the covariance of the observation probability:

- $P(S_t)$. Will be represented as a multinomial, computed on the basis of the sum of expected counts stored in a vector π.
- $P(O_t \mid S_t)$. Due to our conditional independence assumption, the observation probability is written as a product of probabilities $P(O'_t\ O''_t \mid S_t) = P(O'_t \mid S_t)P(O''_t \mid S_t)$. Let us define those probabilities as:

$$P(O'_t \mid [S_t = i]) = \mathbf{G}(O'_t;\ \mu'_i,\ \Sigma') \tag{6.5}$$

and:

$$P(O''_t \mid [S_t = i]) = \begin{cases} \mathbf{U}_{O''_t} & \text{if } O''_t \text{ is not available} \\ \mathbf{G}(O''_t;\ \mu''_i,\ \Sigma'') & \text{otherwise} \end{cases} \tag{6.6}$$

where $\mathbf{U}_{O''_t}$ is a uniform distribution over the goal domain, μ'_i and μ''_i are the mean values of the current and goal components for discrete state i; and Σ' and Σ'' are the respective values of the covariance matrix for all the states.

By noting that $P(O_t \mid S_t)$ is either a product of Gaussians, or a product of a constant and a Gaussian, we may it write this probability as a single Gaussian:

$$P(O_t \mid [S_t = i]) = \frac{1}{Z}\mathbf{G}(O_t;\ \mu_i,\ \Sigma) \tag{6.7}$$

where $\mu_i = [\mu_i', \mu_i'']$, and Σ is a a block diagonal matrix[3] having the form:

$$\Sigma = \begin{bmatrix} \Sigma' & 0 \\ 0 & \Sigma'' \end{bmatrix} \tag{6.8}$$

and Z is a normalization variable, which enables computation of the uniform on the goal component using the same Gaussian representation. Since during prediction the intended part of the augmented observation is not available, this is done by setting[4] $O_t'' = 0$.

- $P(S_t \mid S_{t-1})$. Finally, transition probabilities are stored A in the normal way for GHMMs.

6.4 Prediction

We have not yet discussed prediction, which can be accomplished using the same algorithms that are used for standard HMMs, without interfering with learning. This is possible because learning takes place immediately after an observed trajectory has finished, and thus, it does not affect prediction in any way. For our particular case, we have chosen to apply exact inference:

For every new observation, the current belief state for the object is re-estimated using:

$$P(S_t \mid O_{1:t}) = \frac{1}{Z} P(O_t|S_t) \sum_{S_{t-1}} [P(S_t|S_{t-1})P(S_{t-1}|O_{1:t-1})] \tag{6.9}$$

where $P(S_{t-1} \mid O_{1:t-1})$ comes from the state estimation for the previous time step (or from the state prior, in the case of the first observation in a sequence). Then, prediction is performed by propagating this estimate H time steps ahead into the future using:

$$P(S_{t+H} \mid O_{1:t}) = \sum_{S_{t+H-1}} P(S_{t+H}|S_{t+H-1})P(S_{t+H-1}|O1:t) \tag{6.10}$$

Sometimes, we are interested in knowing the probability of the continuous state probability distribution, as opposed to the discrete space, shown above. Since in our case observations are expressed in terms of the state variable, the continuous state probability can be approximated by the probability that a given state is *observed* in the future, which may be computed from the predicted discrete state as follows:

---

[3] A block diagonal matrix is a square diagonal matrix in which the diagonal elements are square matrices of any size and the off-diagonal elements are zero.

[4] It is easy to show that this is equivalent to a multiplication by a constant, and – when normalized – becomes effectively equivalent to the uniform on Eq. (6.6).

$$P(O_{t+H} \mid O_{1:t}) = \frac{1}{Z} \sum_{S_{t+H}} P(S_{t+H} \mid O_{1:t}) P(O_{t+H} \mid S_{t+H}) \tag{6.11}$$

6.5 Structure and Parameter Learning

By introducing the intended goal into the state variable, we have obtained a state variable which is a good approximation to a *complete* state[5]. This makes it possible to apply our learning algorithm 5.4 in a straightforward fashion.

Thus, besides selecting the algorithm's parameters, the only additional task that should be performed is to preprocess trajectory data in order to obtain the augmented observation sequences $\bar{O}_{1:T}$ described in Sec. 6.2. This is similar to a supervised learning algorithm, where input data contains a label that identifies the motion pattern that has been executed by the object.

6.6 Learning Example: A Unidimensional Environment

To gain a better understanding of the effect of including the intended state in our model, we will discuss in this section the example of a simple one dimensional environment in which objects may only move from a point A ($x = -5\,\text{m}$) to a point B ($x = 5\,\text{m}$) or vice versa as depicted in Figure 6.2. We assume that the probability of observing motion in one sense or the other is the same. Furthermore, we consider that objects move at $1\,{}^{\text{m}}\!/_{\text{s}}$ and observations about the environment are sampled every second.

Fig. 6.2 Example unidimensional environment. Objects move from A to B (blue) or from B to A (green).

6.6.1 Defining the State

It seems natural to start by defining the meaning of the state variable. It makes sense to say that discrete states will represent the object's position. Thus, the extended state will be a two dimensional variable, meaning that for every discrete state i there will be an associated reference vector $w_i = [x_i, x'_i]$, where x_i represents the object's current position and x'_i represents the intended position. Please notice that, for the sake of simplicity, we do not take velocities into account in this example.

---

[5] A state is considered to be *complete* if it is the best predictor of the future , *ie* knowing the state, no prior variables may influence the evolution of future states [cf. Thrun et al., 2005].

6.6.2 Choosing the Parameters of the Learning Algorithm

Since, in this environment, objects move at 1 m/s and observations are sampled every second, we know that objects may move 1 m at every time step. Therefore, we decide that 1 m is an adequate 'step size' for the our problem, and we set the variance to the be the square of the step 'radius' $\Sigma' = [0.5^2]$. With respect to goals, we may be somewhat more flexible, and assume that two trajectories whose last states are within 2 m of each other are indeed attaining the same goal, hence we set the intended state variance to be $\Sigma'' = [1^2]$.

For this particular case, we will set the insertion threshold to be $\tau = 1$, because our assumptions about the discretization size have already been encoded in the covariances. An alternative would be, for example, to increase the threshold and decrease the covariances. We do not want the initially found states to be smoothed, hence, we set the smoothing factor ϵ to zero.

We have now set all the parameters of our algorithm. A key point to highlight is that we have not made any assumptions about the number of goals in the environment, the number of typical trajectories, or the size of the environment. Instead, we have used only our knowledge about the sensors (*ie* sampling rate), the objects dynamics (*ie* velocity) and the semantics of the state space (*ie* unidimensional positions). Therefore, these parameters may be applied to *any* unidimensional environment having the same kind of objects and sensors.

6.6.3 Learned Model

Let us assume that an object goes from A to B, yielding the following observation sequence:

$$O_{1:11} = \{-5, -4, -3, -2, -1, 0, 1, 2, 3, 4, 5\} \tag{6.12}$$

Of course, in most real situations, the observations will not be evenly spaced like these, because of the combined effect of noise, the sampling time, etc. But, we want to keep this example as simple as possible.

Now, we build the extended observation sequence, by appending the last observation in the sequence to every individual observation:

$$\bar{O}_{1:11} = \{[-5,5],[-4,5],[-3,5],[-2,5],[-1,5],[0,5],[1,5],[2,5],[3,5],[4,5],[5,5]\} \tag{6.13}$$

If we plot the points in the sequence, and draw the respective Voronoi regions and Delaunay edges, we will obtain something like Fig. 6.3.

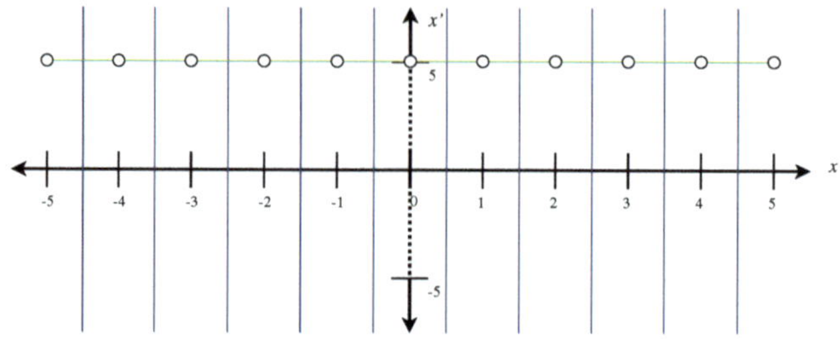

Fig. 6.3 Extended observation sequence plot and Voronoi Diagram. Voronoi region's borders are depicted in blue and Delaunay links in green. x and x' axes are not at the same scale.

As we see, Delaunay edges[6] form a unidimensional manifold, which is a good representation of the structure of the input pattern. Our algorithm will add two transition probabilities for every Delaunay edge, and one self transition by node. So, after processing the extended observation sequence with the learning algorithm, we would obtain the GHMM illustrated in Fig. 6.4.

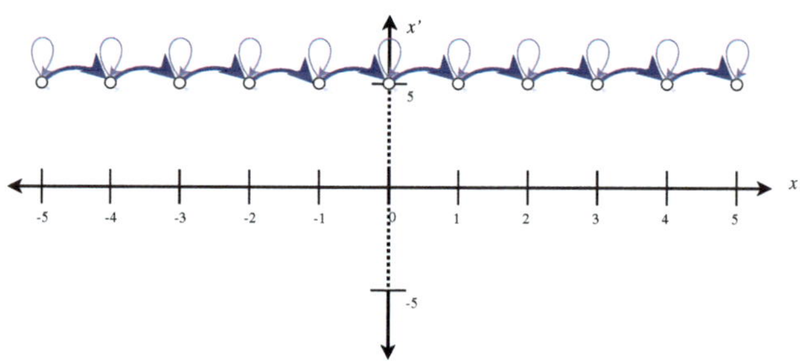

Fig. 6.4 Learned GHMM after processing one kind of observation sequence. The size and color of the arrows represent the probability. x and x' axes are not at the same scale.

In the figure, the transitions which have the strongest probabilities are those which advance the object towards B, while the probabilities of staying in the same state are lesser, and those of going backwards are close to zero. We may be tempted to remove those backward probabilities since, after all, they do not seem to encode any useful information, but we should remember that those edges were added before estimating the parameters, so it would not be possible to not include them without additional knowledge.

[6] In this example, all Voronoi regions are infinite due to the fact that observations are all aligned over the same straight line – we say of this points that they are not in *general position* [cf. Guy, 1989] – hence, all borders are parallel. In normal circumstances, they would be slightly misaligned, so that the borders would cut each other thus defining finite regions.

On the other hand, these low probability transitions only imply a linear increment in the complexity of the model and, due to their low probabilities they do not have a significant effect in the prediction quality. Therefore, we consider that, in general, there is no need to remove these low-probability edges.

Let us assume that now we observe the trajectory of an object which goes from B to A, and the corresponding observation sequence is:

$$O_{1:11} = \{5, 4, 3, 2, 1, 0, -1, -2, -3, -4, -5\} \tag{6.14}$$

We may proceed as above, using our algorithm to update the model. At the end we will obtain the GHMM which is displayed in Fig. 6.5.

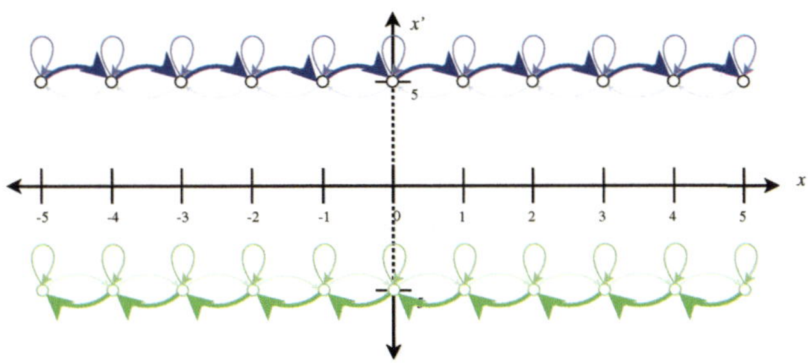

Fig. 6.5 Learned GHMM after processing the two kinds of observation sequences. The size and color of the arrows represent the probability. x and x' axes are not at the same scale.

It is worth noting that, if we had allowed transitions for all the edges in the Delaunay triangulation, there would be edges joining both patterns. They are not there because our algorithm forbids the creation of links which are much longer than the insertion threshold.

As we have shown in this example, our algorithm has obtained a good representation of both motion patterns, without making any assumption about the number of typical trajectories or discrete states in the model. We will not discuss prediction because it is performed using exactly the same procedure than for conventional HMMs/GHMMs, except for the use of extended observations.

6.7 Comparison with Existing HMM Based Approaches

One of the primary goals of the work presented in this book is to develop a lifelong learning algorithm, which is the main difference between our approach and other HMM based motion prediction techniques. However, this is not the only difference, we have also made a different set of hypotheses and modeling choices with respect to most approaches in the literature, which justifies a comparison in terms of these choices.

The two most important differences in our model with respect to other approaches are:

- We model motion patterns in terms of the intended state, while the 'standard' approach[7] is to consider that motion patterns correspond to "typical trajectories".
- While other approaches constrain the structure of a motion pattern to be a chain, our definition allows network-like structures, containing bifurcations and intersections.

These two differences are relevant because they are instrumental in improving two problems related to the use of chains as typical trajectories.

6.7.0.1 Model Redundancy

A first problem that arises from the representation of motion in terms of typical trajectories comes from the fact that structures representing different motion patterns do not share states, which leads to redundant models. A simple example of this situation is illustrated in Fig. 6.6. There are two similar typical trajectories (blue and green), which would be represented as two independent chains (Fig. 6.6(b)) by applying the standard criterion. However, since both typical trajectories end in the same state and most of the intermediate positions are very similar, our algorithm would have produced something similar to Fig. 6.6(c) which is a much more compact representation having the same expressiveness.

(a) Typical Trajectories (b) Chains (c) GHMM

Fig. 6.6 Shared state representation: Two typical trajectories (blue and green, left image) may be represented using two chain-like substructures (center). An alternative (left) is to use a single structure where the states shared by both structures (orange) appear just once.

An interesting situation arises when the occurrence probabilities for the different typical trajectories are not the same. Let us suppose that, in the above's example, the frequency with which the upper trajectory (green) is

[7] There are other goal oriented approaches in the literature (cf. § 3.5.2) but here we are only considering state space models.

observed, is much higher than that of the lower one (blue). If we use chain-like structures, we may represent this by assigning a higher prior probability to the upper chain; unfortunately, most existing approaches assume an uniform prior, thus neglecting this information. By using our approach, the difference in frequencies would be encoded in the outgoing transition probabilities of the split point in Fig. 6.6(c), where the object "chooses" to follow one path or the other.

Despite the advantages of sharing the state, it seems reasonable to ask if the fact of augmenting the state space dimensionality will not lead to an indiscriminate increase in the number of nodes in the Growing Hidden Markov Models (GHMM), thus affecting the scalability of our approach. The answer is no – at least when compared with typical trajectories approaches – because, in those approaches, two trajectories having different final states will, most likely, be represented by two different typical trajectories, thus leading to an increment in the number of nodes in the HMM which is, in the best case, equivalent to our approach.

A weakness that our algorithm shares with other approaches is its assumption that input trajectories have been correctly identified and segmented in a previous step. As we will see in chapter 7, this is not always the case, and the algorithm may end up learning non-existent or fragmentary behaviors.

6.7.0.2 Pattern "Pieces"

When chains are used, transitions between different structures are not allowed, hence, the model is not able to explain motion patterns which are composed of "pieces" of existing patterns, thus ignoring valuable information which is already available. An example of this problem is depicted in Fig. 6.7.

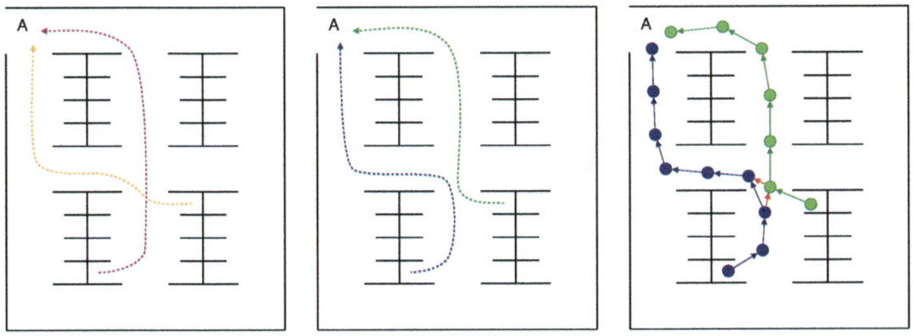

(a) Input trajectories (b) Chain-like (c) GHMM

Fig. 6.7 Example of a situation where pattern "pieces" may be recombined to explain new behavior.

6.7.0.3 Model Semantics

A deeper issue comes from the semantics of this "typical trajectories". They are always defined in terms of similarity, distance or other geometrical or statistical criteria without any reference to the actual *causes* or factors that motivate motion. Therefore, these approaches do not address the crucial question "why are trajectories typical?" which – from our point of view – needs to be answered in order to produce truly generative models of motion. On the other hand, modeling motion causes is difficult in the case of people, because it implies taking the intentional stance towards moving objects. We consider that, although our approach is far from being a satisfactory intentional model, it stands closer to a causal explanation of motion due to the fact that it is based on a rough model of intentions.

6.8 Discussion

In this chapter we presented the application of Growing Hidden Markov Models to learn and predict human motion. Our application is based on the hypothesis that motion is mainly determined by the object's intention to reach a particular state. We have explained how this intention is explicitly modeled by extending the definition of state in order to accommodate not only the current physical state of the object, but also the intended one.

We explained the particularities that should be taken into account to include the extended state into the GHMM formalism. We also provided a basic example of how a concrete problem may be solved using the proposed approach, and illustrated how no assumptions need to be made about the number of discrete states or the number of motion patterns to be learned.

We have compared our approach against other motion prediction techniques based on HMMs, explaining how our proposal leads to improvements in the compactness and the generality of the model.

Recapitulating, the technique we have presented:

- Is able to learn the structure and the parameters incrementally.
- Works on simple parameters which rely on assumptions about the dynamics of the objects and the capacities of the sensors, without requiring prior knowledge on the size of the environment or the number of motion patterns that objects execute in it.
- Produces models which are both rich and conceptually sound.

This concludes the exposition of the theoretic aspects of our work. The next part of this book is dedicated to the experiments we have performed to evaluate our approach.

Part IV
Experiments

Chapter 7
Experimental Data

7.1 Overview

We have conducted thorough experimental validation of the proposed approach. This chapter describes the two data sets which have been used in our experiments:

- *Real data*, obtained through a visual tracking system installed in a parking lot.
- *Synthetic data*, generated with a parking lot simulator.

The presentation of these two data sets is followed by the discussion of the issues we have faced with another visual tracking system, whose data was not usable for our approach and, therefore, constitutes a good illustration of the limitations of approaches that – like ours – strongly rely on having unfailing tracking input.

7.2 Real Data: Leeds Data Set

The first data set was gathered with a visual tracking system[1] installed in an environment featuring a parking lot and a short street section where both pedestrians and vehicles were moving.

Figure 7.1 depicts the process we followed to obtain this data set, its different elements are further discussed below:

- *Camera*: All the video input has been captured by a single camera located high above the environment.
- *Tracking System*: The tracking system proposed by Magee [2004] has been implemented and used to detect and track moving objects in image data. The output of the tracker is a set of trajectories containing position information in image coordinates.

[1] We would like to thank Hannah Dee and the University of Leeds for letting us use this data set.

A.D. Vasquez: Incre. Learn. for Motion Predict. of Pedest. & Vehicles, STAR 64, pp. 97–101.
springerlink.com

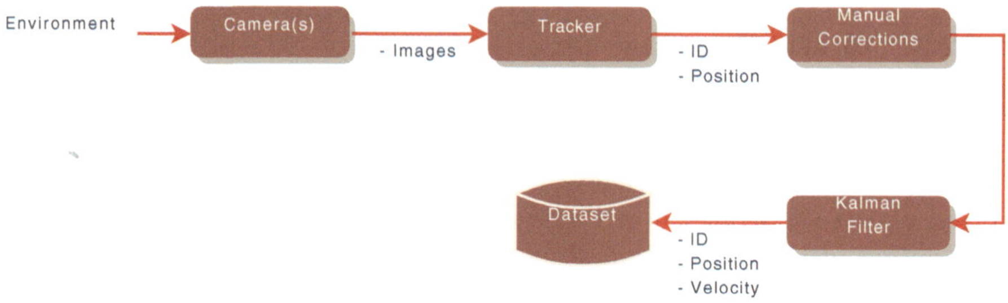

Fig. 7.1 Tracking system overview.

- *Manual correction*: Trajectories have been hand-edited to correct tracking problems. An indicator of the magnitude of this work is that approximately 20% of the trajectories in the data set have been altered in some way, and some have been entirely tracked by hand [Dee, 2005].
- *Kalman Filter*: The observations on this dataset have been sampled at approximately 10 Hz. Since the original data set contains only position information, we have used a Kalman filter to estimate the corresponding velocities.
- *Final Dataset*: The complete data set is depicted in Fig. 7.2, it consists of 269 trajectories.

Fig. 7.2 Leeds data set.

7.3 Synthetic Data: Parking Simulator

Owing to the difficulty of obtaining real data, we have decided to develop a parking lot and trajectory simulator in order to have a more thorough experimental testbed. This allows us to test particular situations without the

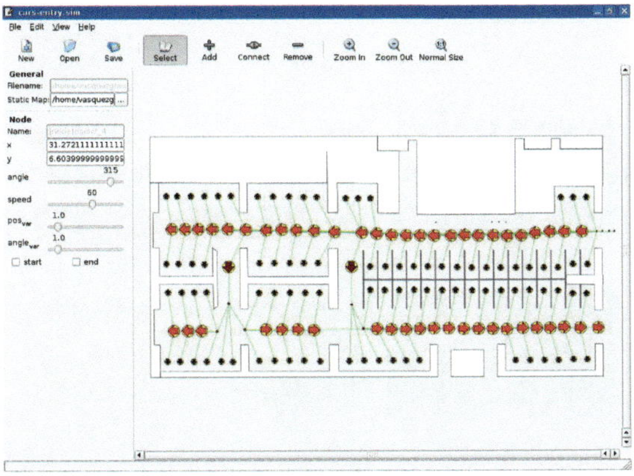

Fig. 7.3 The trajectory simulator. Circled arrows represent control points: arrow size and color correspond to position variance and speed, respectively.

logistic difficulties posed by executing scripted actions in a real environment as the Leeds parking lot. It is also a practical way of generating large amounts of data which are free of the problems that are common to tracking systems.

The simulator is based on a user-defined graph structure (Fig. 7.3). Nodes of the graph represent control points with an associated speed and position variance, *ie* way points in the environment. They also have flags indicating whether they are start, intermediate or end points of a trajectory. Nodes are connected by oriented edges, which indicate that it is possible to go from one node to another.

Once this graph has been defined, the simulator works by choosing at random a start and an end point (according to the corresponding flags), and then obtaining a sequence of nodes by finding the shortest path between them. For every node in the sequence, a point is generated by adding Gaussian noise to the node's position. Finally, the trajectory is generated by applying spline interpolation to the point sequence and sub-sampling it according to the simulated sampling rate and object's speed. As in the case of the Leeds dataset, we have assumed a frame rate of 10 Hz. Figure 7.4 depicts an example trajectory dataset generated by the simulator from the configuration shown in fig. 7.3.

7.4 The INRIA Entry Hall

As we have mentioned at the start of this chapter, we have collected a third data set which presented some issues that, unfortunately, impeded its use in our experiments[2]. The data set was gathered in an indoors environment: the INRIA lab entry hall (Fig. 7.5). The most interesting feature of

[2] As a matter of fact, we conducted some experiments and reported some preliminary results on this data set in Vasquez [2007].

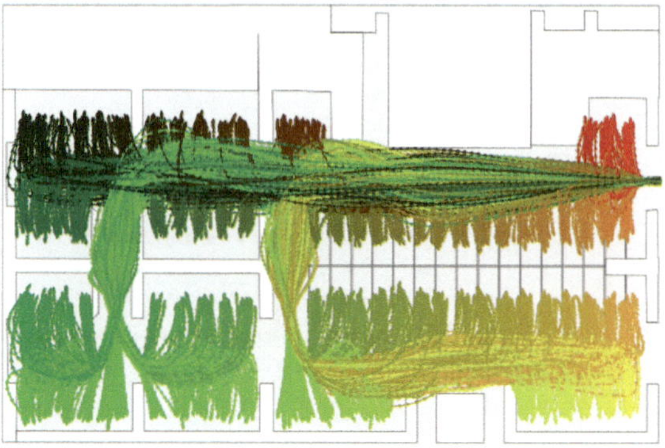

Fig. 7.4 Synthetic parking data set, from the model displayed in fig. 7.3

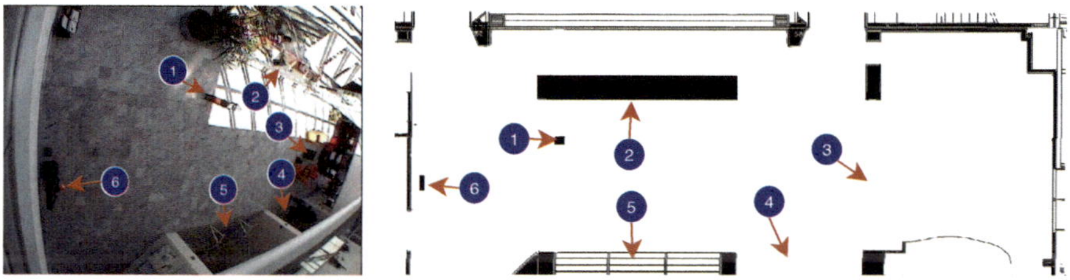

(a) video view (b) floor plan

Fig. 7.5 Inria's main lobby and some landmarks:1) directory post, 2) front desk, 3) coffee area, 4) auditorium entrance, 5) entry stairs, 6) directory post.

this environment is that it is – mostly – an open space and, at the same time, it is not completely unstructured because it has landmarks (*eg* front desk, doors) which condition the way people moves. On the other hand, this environment is still very different from parking lots where the paths – at least for the vehicles – are made explicit by visual cues (*eg* floor markings) and physical obstacles (*eg* sidewalks).

The system used to gather the data was very similar to the one described in section 7.2, except that it used a different tracker proposed by Caporossi et al. [2004]. The main difference, however, is that tracking errors have not been corrected by hand, leading to a big number of "anomalous trajectories" (see Fig. 7.6) in the data set.

Tracking errors related to erroneous data association may split a single trajectory into several shorter ones, or merge the motion of two different objects into a single trajectory. This errors are one of the most important issues faced by automatic motion model learning approaches because they require consistent trajectory data to build meaningful motion models. In practice, most reported experiments use data sets which were corrected by hand [*eg* Hu et al., 2004b, Dee and Hogg, 2004]. A notable exception is the

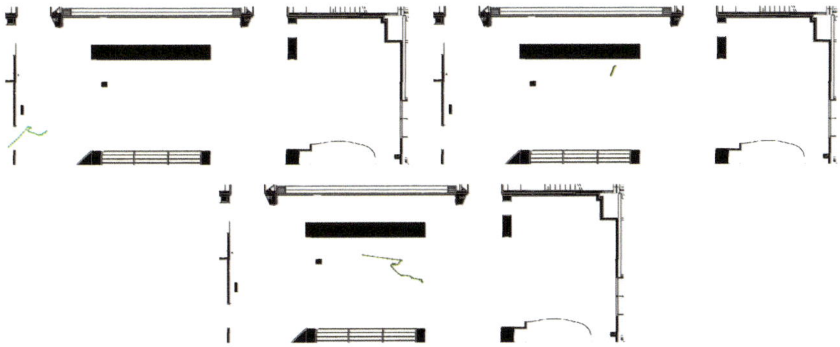

Fig. 7.6 Some examples of anomalous trajectories.

approach of Hu et al. [2006], which filters out low frequency patterns after learning has been performed. However, in the case of lifelong learning where data is processed incrementally, outlier filtering is difficult because newly observed patterns are also statistical outliers that would get filtered out, making it impossible for the system to learn them.

7.5 Discussion

In this chapter we have presented the two data sets we have used for our experimental work, as well as their respective acquisition platforms. For the case of real data, we have illustrated the difficulties that arise from the use of visual trackers to provide input data for typical motion learning approaches. These issues are one of the most important issues that need to be solved in order to have reliable motion learning systems. Robust data association in particular is crucial for this kind of approaches to work.

We will come back to these issues in Chapter 9, where we will discuss what, in our eyes, constitutes an interesting and important research direction: using machine learning techniques to improve tracking and, in particular, the low level motion models used by the tracker's filter.

Chapter 8
Experimental Results

First you guess. Don't laugh, this is the most important step. Then you compute the consequences. Compare the consequences to experience. If it disagrees with experience, the guess is wrong. In that simple statement is the key to science. It doesn't matter how beautiful your guess is or how smart you are or what your name is. If it disagrees with experience, it's wrong.

RICHARD FEYNMAN
From a PBS Show

8.1 Overview

Finding a way to evaluate an approach is often a difficult task, in our case we have chosen to construct our experiments around a few key questions:

1. *Learning performance.* Does learning produce "appropriate" motion models? does it converge?
2. *Prediction accuracy.* How close are predicted states to actual observed motion?
3. *Real-time applicability.* Are learning and prediction fast enough to work at camera rate?
4. *Generality.* Is our approach general enough for general motion prediction? (*eg* cyclic motion, higher-dimensional states).

The objective of this chapter is to present our experimental work on answering these questions. We will start by giving some examples of the application of our approach to the different data sets. Then, we will provide

A.D. Vasquez: Incre. Learn. for Motion Predict. of Pedest. & Vehicles, STAR 64, pp. 103–127.
springerlink.com

numerical measurements of the behavior of our approach as related to our key questions. Next, we will introduce the methodology and present the results of comparing the proposed approach against two other state-of-the-art techniques. Finally, we will close the chapter with a general discussion of the results.

8.2 Qualitative Results

In order to provide the reader with a general understanding of the experiments, this section discusses examples of our approach as applied to both data sets. Technical details such as choosing the algorithm's parameters will be discussed later in the chapter.

8.2.1 Leeds Data

Fig. 8.1 shows an example GHMM structure, obtained after applying our learning algorithm to 100 trajectories of the Leeds data set.

Figures 8.2 and 8.3 illustrate a typical example of the prediction process on the real data set. Fig. 8.3 consists of a set of images arranged in columns and rows. Rows correspond to different values of t.

In each row, the left image shows an actual picture of the parking lot featuring different overlays, as shown in Fig. 8.2(a): a) the current and previous observations, depicted as red dots, b) the current state estimation approximated by a Gaussian indicated with a red ellipse, c) the current goal estimation also approximated by a Gaussian, represented by a golden ellipse, and

Fig. 8.1 Projection on the image of the learned structure after 100 trajectories have been processed (Leeds data set).

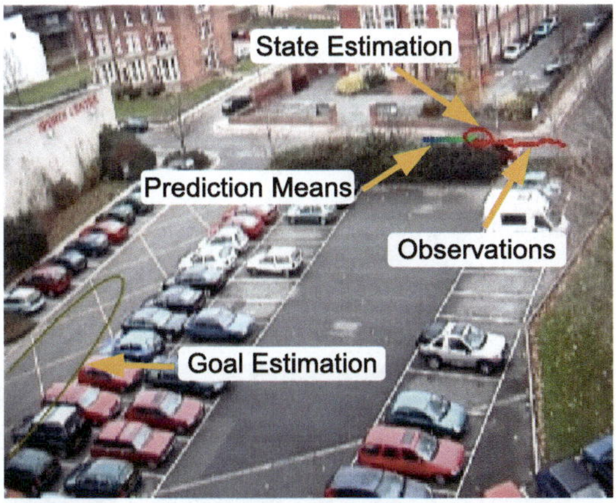

(a) Left Column

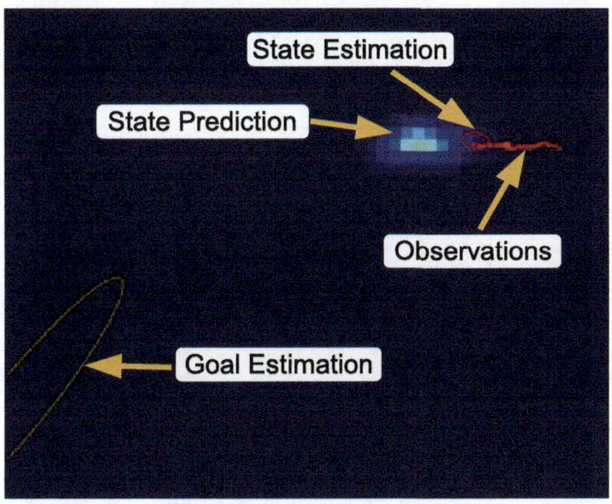

(b) Center Column

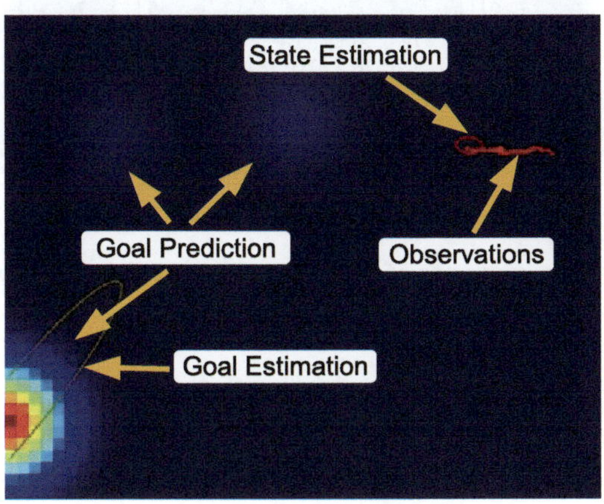

(c) Right Column

Fig. 8.2 Explanation of Fig. 8.3, see § 8.2.1 as well.

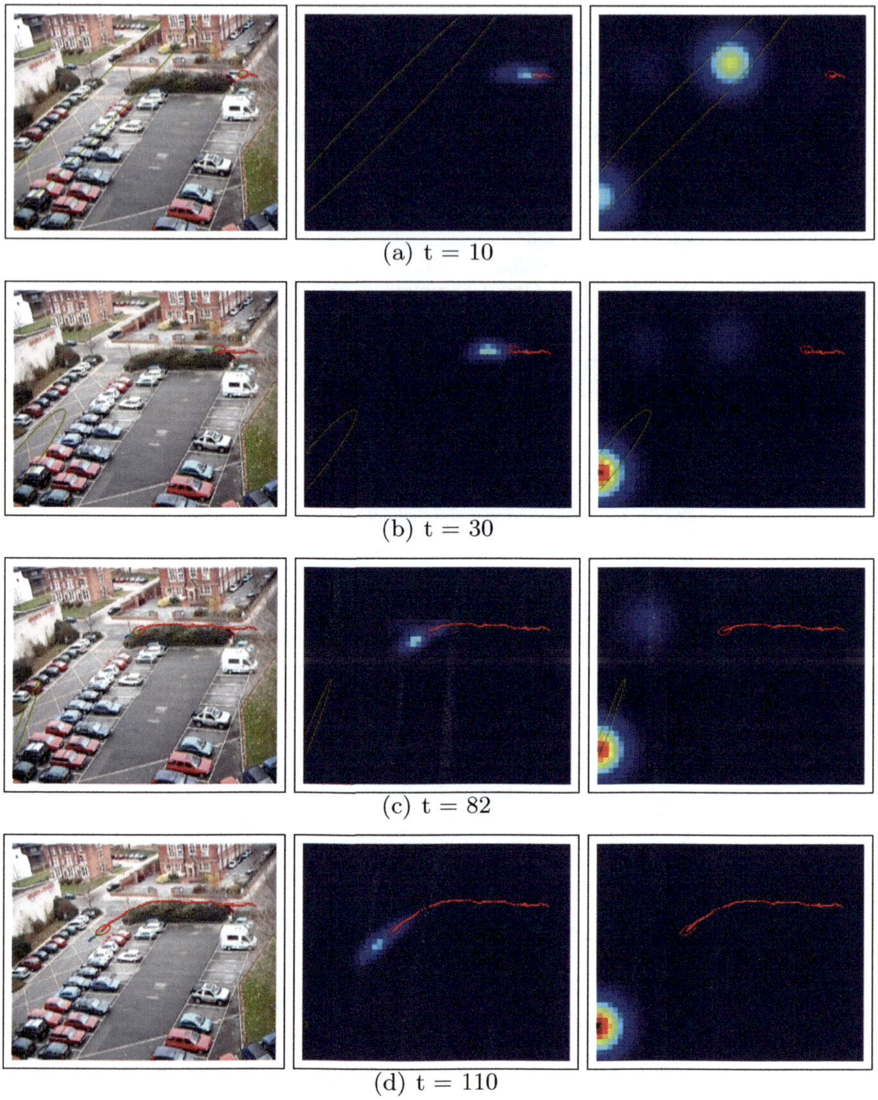

Fig. 8.3 Example of a sequence of predictions for an obstacle moving in the Leeds environment. See § 8.2.1 for details.

d) the mean value of the predicted states for different time horizons going from $H = 1$ to $H = 15$, where H represents the number of time steps to look ahead in the future. These mean values are displayed as points colored from blue (for $H = 1$) to green (for $H = 15$).

The center and right images are 2D projections in the image space of the state and goal estimations. As depicted in Figs. 8.2(b) and 8.2(c) they

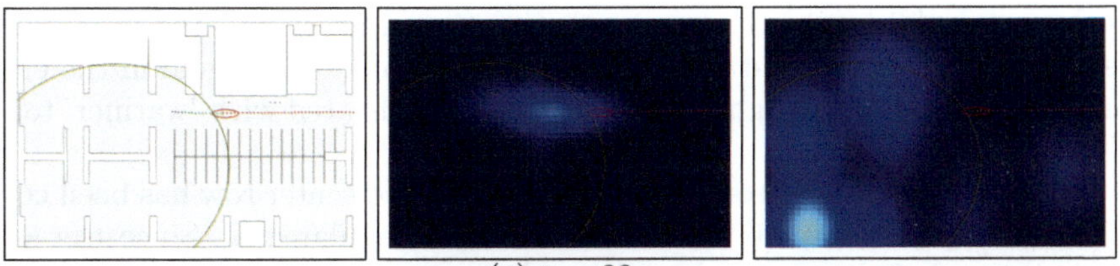

(a) t = 38

(b) t = 49

(c) t = 60

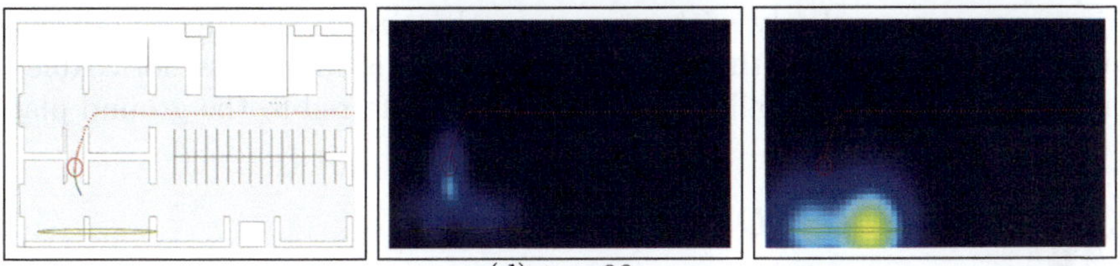

(d) t = 98

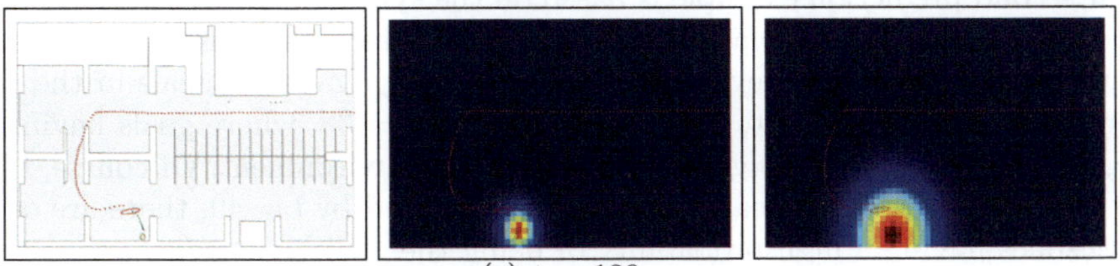

(e) t = 120

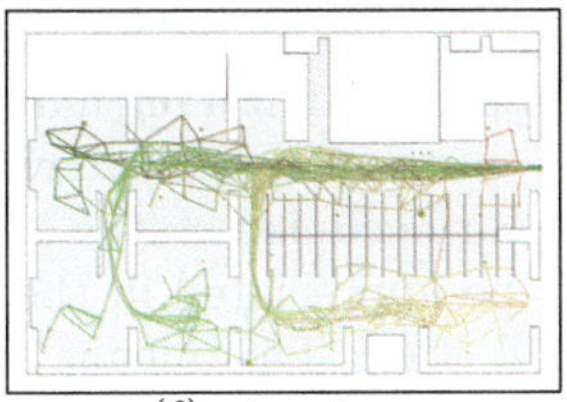

(f) structure

Fig. 8.4 Prediction example (synthetic data).

display – in addition to the previously mentioned overlays – the probability distribution for the predicted position for $H = 15$ and final goal in the environment, respectively. Higher probabilities are indicated with 'warmer' tones (closer to red).

The state prediction probability displayed in the center row has been computed by applying Eq. (6.11) to the cells of a regular grid. Since the augmented state is 6-dimensional, we have chosen to project the probability over the current position plane, thus not showing the predicted goal.

For the right column, we have applied Eq. (6.11) to the cells of a regular grid, much like for the center column, but this time we have projected the probability over the intended position (goal) plane.

An interesting feature of our environment is that it includes two types of moving objects (*ie* pedestrians and vehicles). Since these objects follow different motion patterns, this has considerable influence in the prediction process. For example, for $t = 10$, we may see that there are two highly probable goals. This is interesting because they correspond to a pedestrian's destination (the building entrance) and a vehicle's destination (a lane's end). As the vehicle moves further, it becomes quickly associated with a vehicles' goal and, by $t = 82$ the only two goals with a significant probability correspond to vehicles' destinations.

Also noteworthy is that predicted states at $H = 15$ seem to be considerably close from the current object position. The reason is that, in this data set, objects move very slowly with respect to the size of the image, especially when they are far from the camera. This effect is much less noticeable on simulated data, because all its coordinates are referred to the ground plane.

8.2.2 Simulated Data

Our second prediction example is based on the synthetic data set. The corresponding images are displayed in Fig. 8.4. Something that distinguishes this environment from the previous one is the bigger number of goals in the environment. This is patent, for example, in row $t = 38$, where goals having a significant probability roughly cover half of the environment. Of course, this situation changes as the car continues to move and by $t = 49$, there are only two zones having a high probability of being the goal.

Having so many possible goals also influences the shape of the predicted state probability, which may become quite irregular, as in the case of $t = 49$. This irregularity is also an indicator that a big number of alternative hypotheses are being taken into account, which is unavoidable without additional information such as the free or occupied status of parking places.

Something that may not be appreciated in the images is that predictions are made in terms of the full state. This means that velocity information is also taken into account in order to increase the selectivity of the model.

8.3 Quantitative Results

In this section, we present the numerical results of the experiments we have carried out in order to answer our key questions.

8.3.1 Parameter Selection

For the purpose of studying the sensibility of our approach to parameter changes, we have used six sets of parameters when applying our learning algorithm to the data sets, as described by the following table:

Data Set	Units	Parameter Set	Threshold	σ_{pos}	σ_{vel}	σ_{goal}
Leeds Parking	*pixels*	Guess	4	10	0.9	20
		Low CV	4	7	0.9	10
		High CV	4	12	0.9	20
		Low IT	3	10	0.9	20
		High IT	6	10	0.9	20
		Best	9	7	0.9	20
Synthetic Data	*meters*	Guess	4	2	0.04	5
		Low CV	4	1.5	0.04	3
		High CV	4	3	0.04	6
		Low IT	3	2	0.04	5
		High IT	6	2	0.04	5
		Best	9	1.5	0.04	4

In both cases, we have used spherical covariances for the state variables, hence the table only displays the corresponding standard deviation. In other words, the covariance matrix has the form:

$$\Sigma = \begin{bmatrix} \sigma_{pos}^2 & 0 & 0 & 0 & 0 & 0 \\ 0 & \sigma_{pos}^2 & 0 & 0 & 0 & 0 \\ 0 & 0 & \sigma_{vel}^2 & 0 & 0 & 0 \\ 0 & 0 & 0 & \sigma_{vel}^2 & 0 & 0 \\ 0 & 0 & 0 & 0 & \sigma_{goal}^2 & 0 \\ 0 & 0 & 0 & 0 & 0 & \sigma_{goal}^2 \end{bmatrix}$$

Parameters were fixed as follows: the first parameter set corresponds to an initial guess; the "Low CV" and "High CV" parameters correspond to smaller and bigger covariance values, respectively; the "Low IT" and "High IT" do the same thing for the insertion threshold; the last parameter set is the best one we have found with respect to its parsimony and prediction accuracy, in all cases it was obtained by trial-and-error over several combinations of insertion threshold and covariance values.

8.3.2 Measuring Prediction Accuracy

A common performance metric for probabilistic approaches is the maximum data likelihood or approximations like the BIC (see §4.6). However, for our particular application, this metric has the drawback of not having any geometric interpretation. Intuitively, we would like to know *how far* was the predicted state from the real one. Hence, we have preferred to measure the performance of our algorithm in terms of the average error, computed as the expected distance between the prediction for a time horizon H and the effective observation O_{t+H}.

$$\langle E \rangle = \sum_{i \in \mathcal{S}} P([S_{t+H} = i] \mid O_{1:t}) \|O_{t+H} - \mu_i\|^{1/2} \tag{8.1}$$

for a single time step. This measure may be generalized for a complete data set containing K observation sequences:

$$\langle E \rangle = \frac{1}{K} \sum_{k=1}^{K} \frac{1}{T^k - H} \sum_{t=1}^{T^k - H} \sum_{i \in \mathcal{S}} P([S_{t+H} = i] \mid O^k_{1:t}) \|O^k_{t+H} - \mu_i\|^{1/2} \tag{8.2}$$

It is worth noting that, as opposed to the standard approach in machine learning of conducting tests using a “learning” and a “testing” data sets, the experiments we present here used a single data set. The reason is that learning takes place after prediction and, in consequence, there is no need for such separation: every observation sequence is “unknown” when prediction takes place.

8.3.3 Leeds Parking Data

Figure 8.5 illustrates the behavior of our algorithm on the 269 trajectories of the Leeds data set for the different parameter groups listed in § 8.3.1. The top graph displays the evolution of the average mean error with respect to the number of trajectories that have been processed. The graph on the bottom shows the evolution of the model’s size, measured as the total number of edges.

A first thing to notice is that higher covariance values lead to models having less edges, at the cost of an increased error measure. This is a natural consequence of the fact that the “grain” of the discretization is in part determined by the covariance. Conversely, lower values for this parameter reduce the error measure, but increase the model’s size. It also appears that changes in the covariance are roughly proportional to changes in the number of edges.

However, the insertion threshold behaves less predictably, in particular in the case of higher values, which does not seem to affect the error measure considerably. This suggests that the model’s size may be decreased without sacrificing accuracy by using relatively high insertion thresholds. In fact,

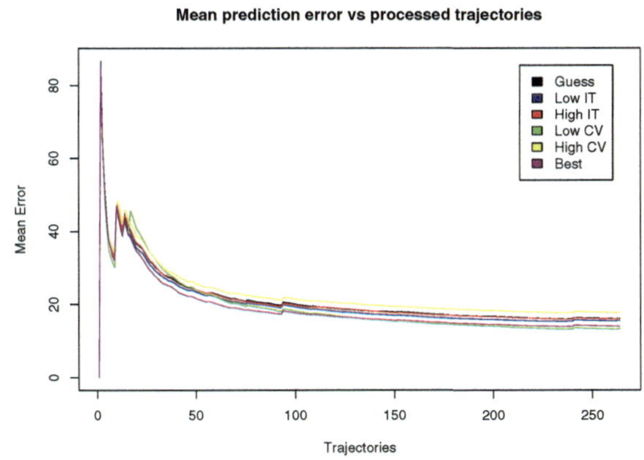

(a) mean error

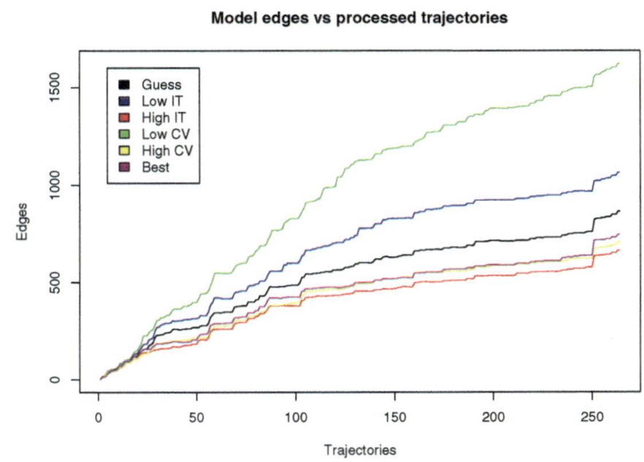

(b) model edges

Fig. 8.5 Parameter sensibility (Leeds data).

the best parameter set has been obtained by simply increasing the insertion threshold of the low covariance parameter set, obtaining a comparable performance while dividing the model size by two.

The rest of our analysis for this data set will focus on results obtained with the best parameters. Fig. 8.6(a) shows the evolution of both the model's size and the average error as a function of the number of processed trajectories. There seems to be an overall convergence process combined with sharp step-like increases in both model size and average error (see for example the graph at 100, 160 and 250 processed trajectories). The explanation for these steps is that they correspond to motion patterns that have been observed for the first time and, thus, require the insertion of several new discrete states in the model in order to be described.

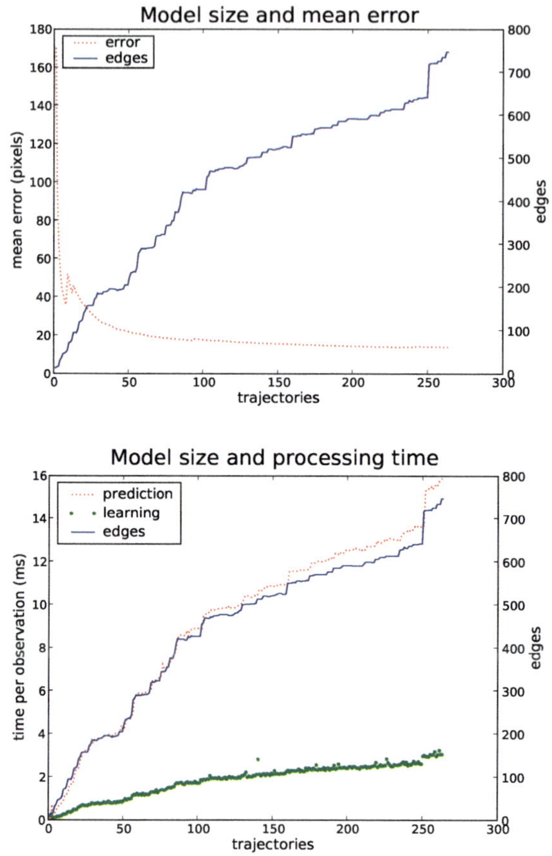

Fig. 8.6 Error and computation times for the best parameters (Leeds data).

Fig. 8.6(b) shows a plot of the time taken by prediction and learning against the number of processed trajectories, the number of edges is also plotted as a reference. Times are given per-observation, hence, in the case of learning, they should be multiplied by the length of the observation sequence. As it may be expected, the increase in processing time seems to be proportional to the model's growth. Regarding real-time applicability, it is worth noting that the time required by the learning algorithm to process a single observation is always below 5 ms meaning that a ten second trajectory would require slightly more than one second to be learned. Thus, the algorithm is well adapted for on-line usage in environments with moderate pedestrian traffic, where it is likely to observe less than one new trajectory per second.

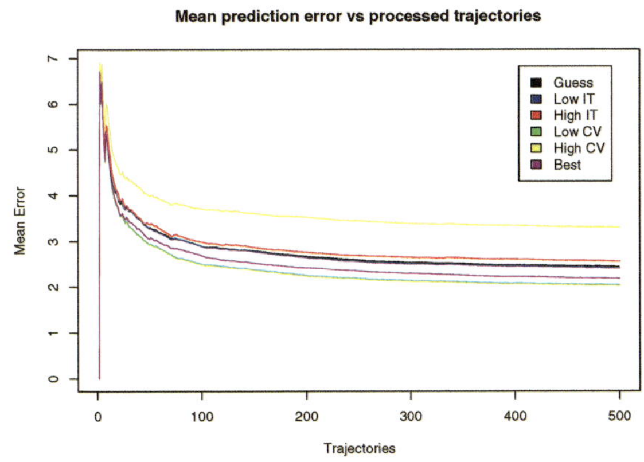

(a) mean error

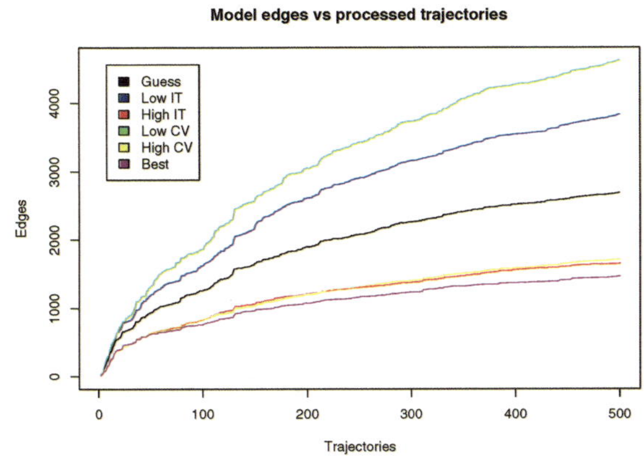

(b) model edges

Fig. 8.7 Parameter sensibility (synthetic data)

8.3.4 Synthetic Parking Data

The second data set has many more possible destinations, with about 90 parking places. As we will see, our algorithm performed surprisingly well, taking into account this added complexity. In the parameter sensibility graphs in fig. 8.7 we see that the algorithm behaves mostly as in the previous case, except that here the best parameter set has yielded the smallest model – by a small margin – while having the second best average error.

Given the size and features of this data set, it was surprising to find that the error evolution and model growth (fig. 8.8(a)) performed that well, with a final number of edges below 1500 despite the large number of possible goals. We believe that this reduced model size is mostly due to the environment's

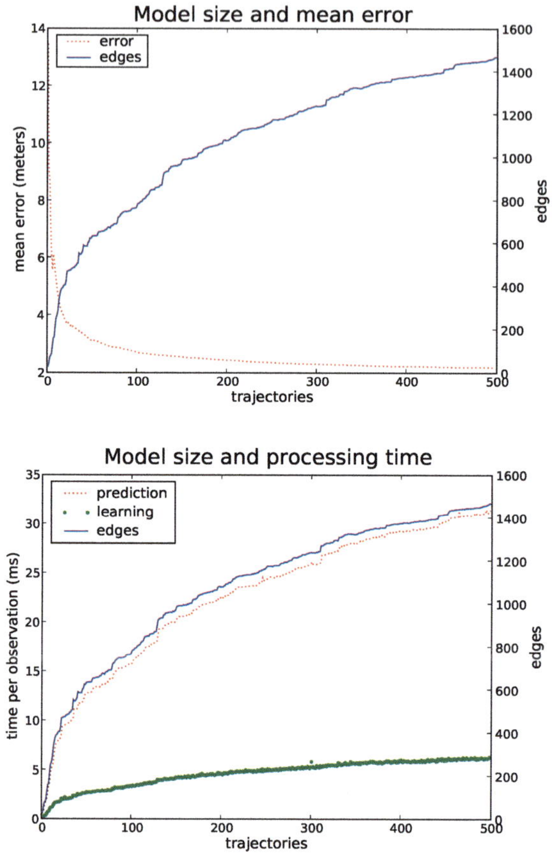

Fig. 8.8 Error and computation times for the best parameters (synthetic data).

structure, where trajectories leading to goals in the same area tend to share a considerable number of states.

As for the previous data set, time performance was very good, with prediction times below 35 ms, and learning times under 5 ms per observation after 500 processed trajectories. These values show, once again, the real-time applicability of our approach in areas with moderate traffic.

8.4 Modeling Cyclic Motion

A question that has been often posed to us is how our approach deals with motion patterns containing cycles. As an example, let us imagine a 800 m track running event, where contestants have to go twice around the race-track in order to finish the race. It is evident that the situations at 395 and 795 meters are very different, in the first one, the competitor will continue to run

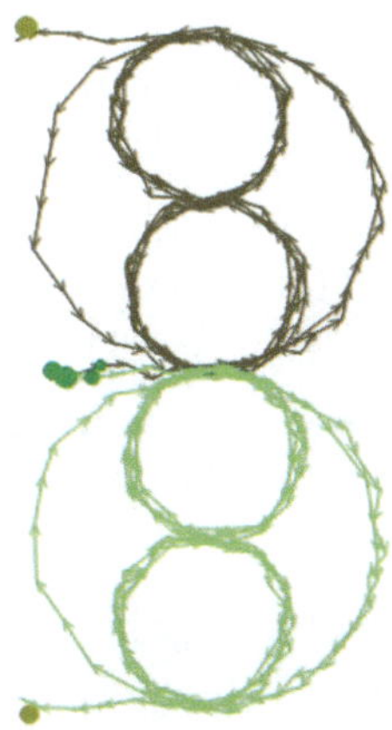

Fig. 8.9 Examples of cyclic trajectories. One trajectory per motion pattern is shown, the upper (gray) and lower (green) trajectories corresponds to left and right turns, respectively. See § 8.4 for further details.

for another turn, while in the second one it will stop after five meters. Since our motion prediction approach as it is defined, only distinguishes discrete states in terms of their position and speed, it is not able to deal with this situation, resulting in equiprobable predictions of stopping and continuing at both 395 and 795 meters.

Although this example may seem contrived, this situation arises frequently for other kinds of motion, for example in the case of human gestures. Here, we propose a simple solution to this problem: including the time variable t in observations. To illustrate this we have prepared a synthetic data set consisting of two symmetric motion patterns (see Fig. 8.9) where the object describes one and a half big circles, interleaved with two smaller cycles at the top and bottom. The only difference between both patterns is that one always turns to the left and the other one to the right.

Of course, including the *time* variable in observations implies augmenting the observation covariance matrix, to include the time variable. Without going into the parameter details, Fig. 8.10 illustrates our algorithm working on this data set.

Notice how object's positions for $t = 20$, $t = 60$ and $t = 100$ are very similar. Nevertheless predictions are clearly different. At $t = 20$ the object has not even turned, hence, the estimated probability for both goals practically the same. After one turn, at $t = 20$, the algorithm has identified the left turning motion pattern, and it predicts that the object will still perform one more small turn. Finally, at $t = 100$ the algorithm recognizes that the object has performed two small turns and thus predicts that the object will continue to follow the big one.

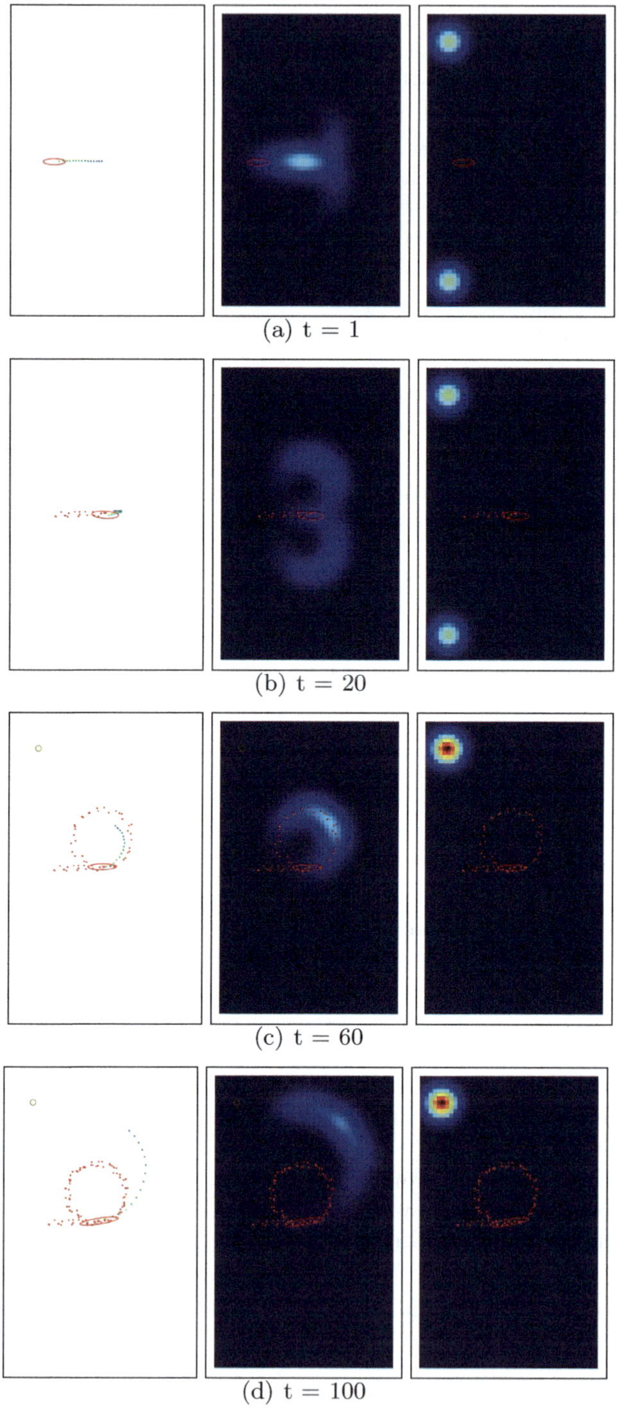

Fig. 8.10 Prediction example for the cycle data set.

As our example shows, by simply redefining observations, our algorithm is able to deal with cyclic motion patterns without further modifications. Moreover, moderated temporal misalignments may be easily handled by modifying the covariance matrix.

8.5 Comparison with Other Approaches

This section provides a quantitative comparison of our approach against two other recent techniques. First, we introduce those approaches, then, we describe the proposed performance measures, finally we discuss the results we obtained with both real and simulated data.

8.5.1 Compared Approaches

We have implemented two other approaches in order to compare them to the proposed technique. These approaches have been selected among the many different approaches discussed in chapter 3, on the basis of three criteria:

- Unsupervised learning. Since our approach learns from unlabeled data, we are interested in comparing it against similar approaches.
- Structure learning. The compared approaches should be able to estimate the size and structure of the learned model, not only the parameter values.
- Suitability for prediction. Not all the existing approaches are suited for motion prediction, which, from our point of view, requires at least two conditions: the approach should model time – at least implicitly – and it should be able to produce multi-modal predictions.

From the approaches that verify those criteria, we have selected an HMM-based approach, which applies the expectation-maximization algorithm for learning, and a second approach which is based on hierarchical fuzzy K-means clustering.

8.5.1.1 Expectation-Maximization Clustering

This approach, proposed by Bennewitz et al. [2005] uses the expectation-maximization (EM) algorithm to cluster together trajectories into trajectory prototypes, which are then transformed into HMMs with the same semantics than in our approach, but limited to chain-like structures where the only valid transitions out of a state go to itself and to the next state in the chain.

We will discuss here the overall structure of the clustering algorithm, the interested reader is referred to the original paper for further details:

1. Initialize K clusters from randomly chosen observation sequences.
2. Run optimization.

3. Compute model score.
4. If the score may be improved by adding the worse scored trajectory as a cluster, add it and go to step 2.
5. If the score may be improved by removing one of the two more similar clusters, delete it and go to step 2, else stop.

Step two is the most important, it estimates the cluster parameters by iterating through two steps: a) for every observation sequence and cluster, compute the expected membership, *ie* the likelihood that the observation sequence belongs to that cluster, and b) reestimate cluster representations as an average of observation sequences, weighted by the corresponding likelihoods that were computed on the previous step. The process is stopped when parameter changes between two consecutive iterations are negligible.

Since the EM algorithm requires the number of clusters to be known *a priori*, it is necessary to estimate it somehow. This is the purpose of steps 3 to 5, which score the model using the BIC (see § 4.6) and then try to improve that score by adding or removing clusters. It is worth noting that our implementation differs slightly from the original paper, which worked with predefined insertion and deletion thresholds for steps four and five. Since we have found it difficult to fix satisfactory values for these thresholds, we have preferred to always try to increase and decrease the cluster number in order to see if the score is improved.

Once the clusters are found, they are converted into HMMs by applying prior knowledge about how objects move. HMM states are uniformly spaced along the found clusters, and transition probabilities are fixed according to average object velocities. A final note about this approach is that it does not use velocity information on the observations – although the authors mention the possibility to extend the approach to use such information.

8.5.1.2 Hierarchical Fuzzy K-Means Clustering

The second approach we have implemented [Hu et al., 2006] is based on the fuzzy K-means clustering algorithm (HFKM). It has the same overall structure than expectation-maximization, but uses different expressions to compute the expected memberships and cluster representations in step 2. Also, a different measure – the Tightness and Separation Criterion [Xie and Beni, 1991] – is used to find the optimal number of clusters on steps 3 through 5.

The approach works in a hierarchical manner: a first clustering is performed on coarsely subsampled observation sequences using only pose information. Then, for every obtained cluster, the belonging observation sequences are clustered again, this time without sub-sampling and taking into account the object's velocities.

In this case, our implementation differs again from the original paper: after conducting an initial round of tests which produced unsatisfactory clustering results, we concluded that there is a typo in cluster representation computation formula that appears in the paper (see appendix B). After the corresponding modifications, the results improved significantly. A final difference with the EM approach is that, after the final clustering step, "anomalous" trajectories are filtered out by deleting clusters that have less than a given number of trajectories. In our experiments we have fixed this threshold to three.

8.5.2 Comparison Criteria

Comparing heterogeneous techniques – even if they solve a common problem – is a difficult task. Often their theoretical bases are too different, making it difficult to evaluate them fairly. Here we propose two measures that, despite their simplicity, still provide useful indicators about the accuracy and the parsimony of the models produced by the different approaches.

8.5.2.1 Measuring Model Size

The structure of the transition matrix is the fundamental complexity factor for inference on Hidden Markov Models, a sparse transition matrix implies much lower computation cost than a dense one. Therefore, we measure the size of the model by the number of edges in the transition graph. For GHMMs this is the sum of the number of neighbors for every discrete state in the model. In the case of EM and HFKM this is the sum of the transitions for all clusters, which for a single cluster is equal to the number of points in the cluster minus one, plus the number of points, to account for self-transitions.

8.5.2.2 Measuring Prediction Accuracy

To evaluate prediction accuracy, the average expected prediction error (Eq. 8.2) has been computed from a test data set containing K observation sequences. It should be noted that the case of the HFKM algorithm is particular, since the algorithm only outputs the probability of following a given motion pattern and is not able to produce predictions at the state level. In order to compare approaches, we have assumed a deterministic transition probability where all the probability mass for the nth time step is concentrated in the nth point of a cluster:

$$P([S_{t+H} = i] \mid O_{1:t}, \phi_j) = \begin{cases} P(O_{1:t}|\phi_j) & \text{if } i \text{ is the } t+H\text{-th element of cluster } j \\ 0 & \text{otherwise} \end{cases} \tag{8.3}$$

where $P(O_{1:t}|\phi_j)$ is the probability[1] that the observation sequence $O_{1:t}$ belongs to cluster j.

[1] See [Hu et al., 2006] for a definition of this probability.

8.5.3 Leeds Data

In order to compare the performance on the Leeds environment, we proceeded by dividing the data into a learning data set (200 sequences) and a test data set (60 sequences). To evaluate how the model's size and accuracy evolve with respect to the size of the learning data set, we performed five experiments, giving 40, 80, 120, 160 and 200 input trajectories to the learning algorithms. In the case of the GHMM and to have a fair comparison, learning has been performed on those sequences prior to prediction. However, it should be noted that this is by no means a requirement for our approach since it is designed to learn and predict in a continuous fashion.

The parameters we have chosen for every algorithm are shown in table 8.1. In the case of our approach, we have used the *best* parameter set (*cf* 8.3.1), the parameters of the other approaches were chosen by making an educated guess and then refining them by trial and error.

Table 8.1 Parameters for Leeds data.

Algorithm	Parameters
EM	$\sigma = 7, K_0 = 15$
HFKM	$sampling\_step = 25, K_0 = 15$
GHMM	$\sigma^2_{pos} = 49, \sigma^2_{vel} = 0.8, \sigma^2_{goal} = 400, \tau = 9$

The meanings of the parameters that appear in the table and are not defined in the text are the following: K_0 is the initial number of clusters. For HFKM *sampling_step* means that in the first clustering step, learning sequences will be sampled by taking one out of *sampling_step* observations on them.

8.5.3.1 Comparing Prediction Accuracy

Figure 8.11 shows the average prediction error as a function of the total number of trajectories in the learning data set. For every batch, full learning is performed and then the expected error is computed with Eq. 8.2 using the test data set as input.

As it can be seen, the average error is much lower for GHMMs than for the other two approaches. Perhaps more surprising: while for GHMMs, the error seems to decrease as more trajectories are used for learning, this is not the case for the other two algorithms. At first, we thought that this was due to initialization differences between runs, where different trajectories were used to initialize the clusters in each case. In order to discard this possibility, we have used the same initialization for all different runs without significant difference, as the figure shows. Our current explanation is that the fact of using a global optimization scheme makes the algorithm very sensitive to

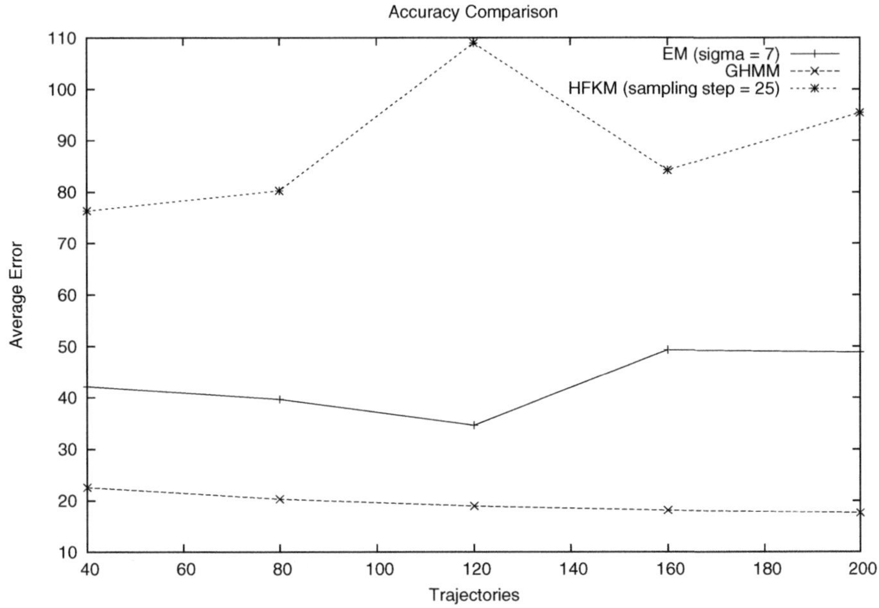

Fig. 8.11 Leeds data: prediction accuracy.

initial conditions, leading to very different clusterings depending on input data, sometimes with suboptimal results.

Another important factor, at least in the case of HFKM, seems to be that the model is "saturated" and is not able to perform further generalization on the basis of additional data. This leads us to what, in our eyes, is an important drawback of HFKM: it lacks some kind of scale parameter (like the covariance in EM and GHMM) to trade off a better accuracy for the cost of having a more complex model. This will become clearer in the model size comparison.

8.5.3.2 Comparing Model Size

The growth on model size with respect to the number of trajectories in the learning data set is displayed on fig. 8.12. As it can be seen in the figure, the size of the GHMM models is negligible with respect to the other two approaches. On the other hand, the model growth tends to stabilize for both GHMM and HFKM, while in the case of the EM approach, it jumps suddenly for 160 trajectories. As mentioned in § 8.5.3.1, this seems to indicate that both GHMM and HFKM have converged, but the behavior of EM is more difficult to explain.

The jump is explained by a number of "atypical" trajectories, that appear between positions 120 and 160 in the learning data set. These trajectories

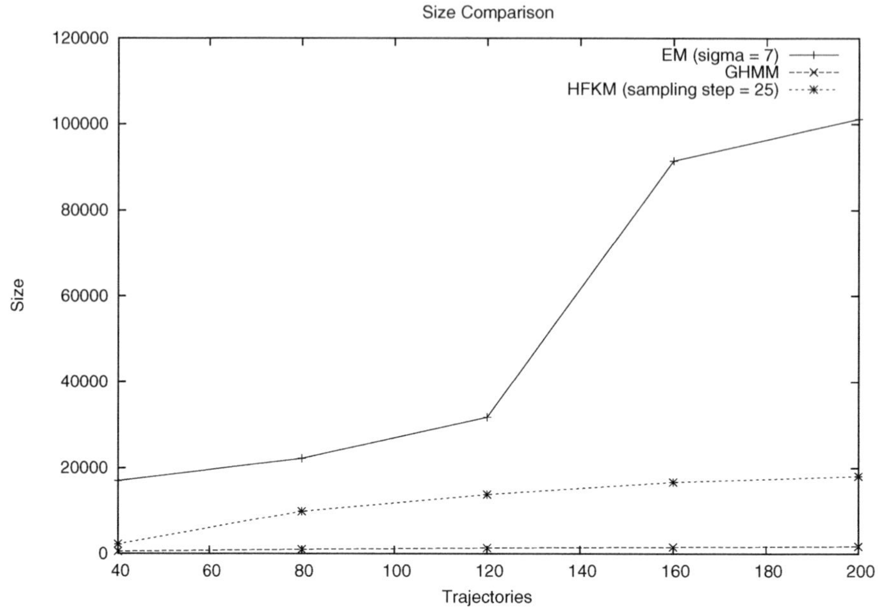

Fig. 8.12 Leeds data: model size.

force the creation of several clusters in the EM model that do not appear in the HFKM approach due to its cluster size threshold. In the case of GHMM, these trajectories also lead to a sudden growth of the model, but this is more difficult to perceive because of its small size.

Finally, in Fig. 8.13, we plot the model size against the prediction error. This illustrates more explicitly the fact that, for the HFKM and EM algorithms, an increase in the model size does not necessarily lead to an improvement in the accuracy. As we discuss in the following subsection, we have explored this further using simulated data.

8.5.4 Synthetic Parking Data

In the case of the simulated parking environment, we have proceeded in analogous fashion to what we have done with the Leeds data set, but this time, the learning data set contains 1000 sequences while the test data set contains 100. The learning batches for this data set contain 200, 400, 600, 800 and 1000 trajectories, respectively.

The parameters for every algorithm are shown in Table 8.2. This time, however, we have included two different sets of parameters per algorithm in order to show their influence on the model performance. The parameters have also been obtained by trial and error.

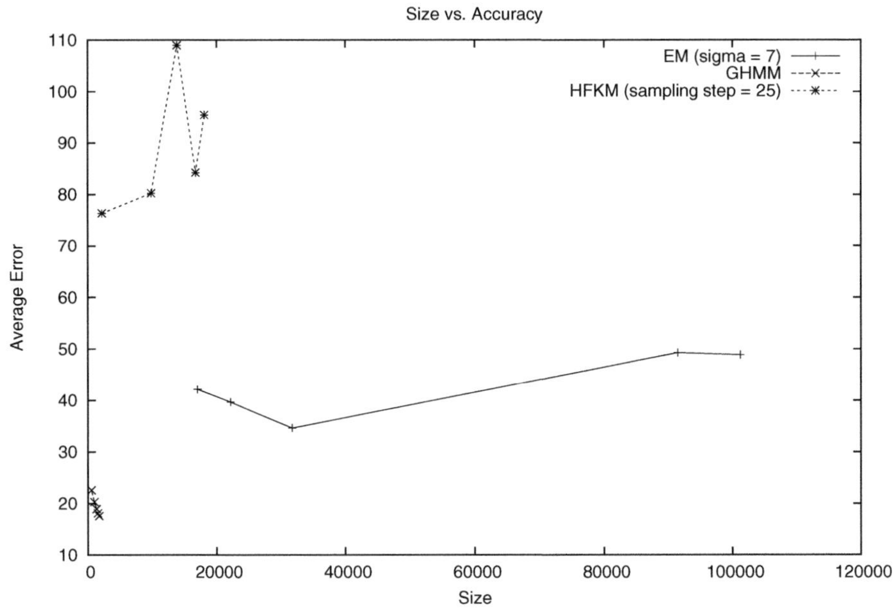

Fig. 8.13 Leeds data: model size vs. prediction accuracy.

Table 8.2 Parameters for synthetic data.

Algorithm	Parameters
EM (sigma = 0.5)	$\sigma = 0.5, K_0 = 15$
EM (sigma = 1)	$\sigma = 1, K_0 = 15$
HFKM (sampling step = 3)	$sampling\_step = 3, K_0 = 15$
HFKM (sampling step = 25)	$sampling\_step = 25, K_0 = 15$
GHMM (cov1)	$\sigma^2_{pos} = 2.25, \sigma^2_{vel} = 0.04, \sigma^2_{goal} = 16, \tau = 9$
GHMM (cov2)	$\sigma^2_{pos} = 1, \sigma^2_{vel} = 0.04, \sigma^2_{goal} = 16, \tau = 9$

8.5.4.1 Comparing Prediction Accuracy

In this data set, the accuracy of the GHMM approach is again better than that of the compared approaches, but this time the difference is not as substantial as for Leeds data. Moreover, the behavior of the other approaches – with the exception of HFKM with a sampling step of three – is this time much closer to that of the GHMM approach. This is explained by the absence of atypical trajectories in the learning and test datasets, together with the existence of many similar example sequences per motion pattern.

We have not found the reason why HFKM with a sampling step of 3 behaves differently, our best explanation at this moment is that, as the number of points in a trajectory increases, the clustering starts to smooth out the

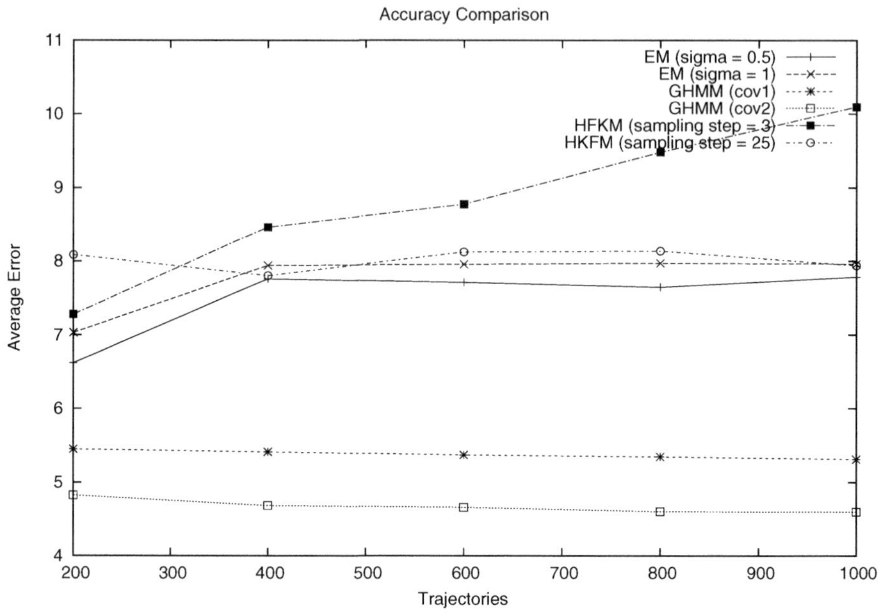

Fig. 8.14 Synthetic data: prediction accuracy.

distinctive part of every trajectory and to produce less clusters, which fail to accurately represent the underlying behaviors.

8.5.4.2 Comparing Model Size

Regarding the size of the model (fig. 8.15), the results obtained with the simulated data set are very different from the Leeds data set. In this case the size of the models obtained with both GHMM parameterizations are similar to the other approaches, with the exception of EM (sigma = 1). On the other hand, once again, our approach is the only case where the model size growth is monotonic, while for the other approaches it tends to oscillate. In fact for both HFKM parameterizations, the model size seems to decrease as the learning data set becomes larger.

By plotting the model size versus the accuracy (Fig. 8.16) we observe again that, with the exception of GHMMs, for the same parametrization, an increase in the model size is not translated directly into an improvement in the accuracy. It also shows that, between models of comparable size, those obtained with our approach exhibit consistently and significantly better accuracy.

This is also a good opportunity to recapitulate on the choice of the model parameters. It is probable that, at least in the case of EM, reducing the covariance matrix could improve the accuracy to make it closer to what is

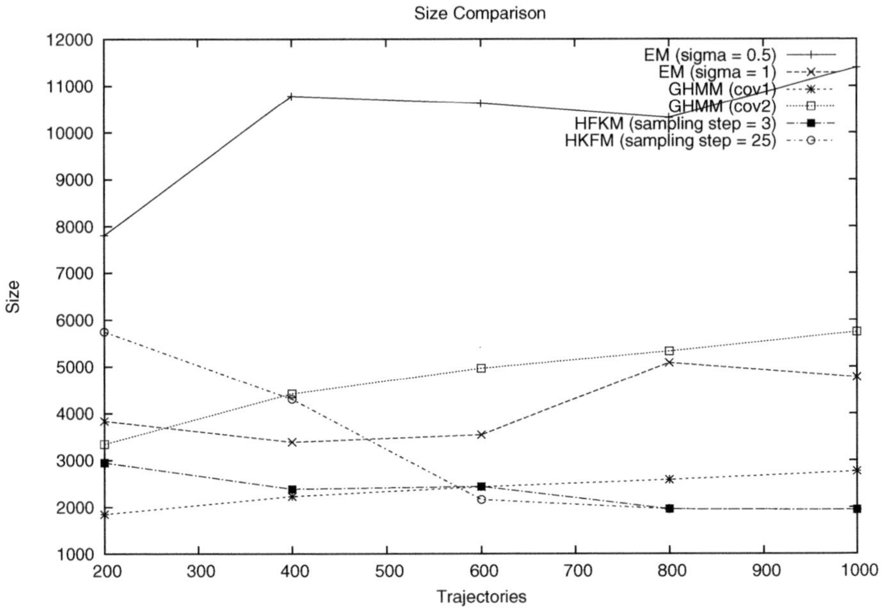

Fig. 8.15 Synthetic data: model size.

obtained with GHMM, nevertheless, this would imply a huge increase in the model size, which would not be comparable to that of the GHMM model. This seems to indicate that our model generalizes better, and that, in the case of EM, the model is probably overfitting data.

8.5.5 Evaluation Conclusion

Despite the good results that we have obtained so far, we consider that much more experimental work needs to be done before arriving at definitive conclusions. We consider at this moment that our approach performs better than the similar EM approach mainly because of two factors: a) it has a more expressive and compact structure and b) it learns the parameters of the observation and transition probabilities from data instead of relying in *a priori* knowledge. We consider the first factor to be more important than the second; it would be interesting, however, to apply HMM parameter learning to the EM approach after trajectory clustering, to have a quantitative evaluation of the relative importance of both factors. It would be also interesting to do something similar in the case of HFKM, where an HMM is built from the obtained clusters.

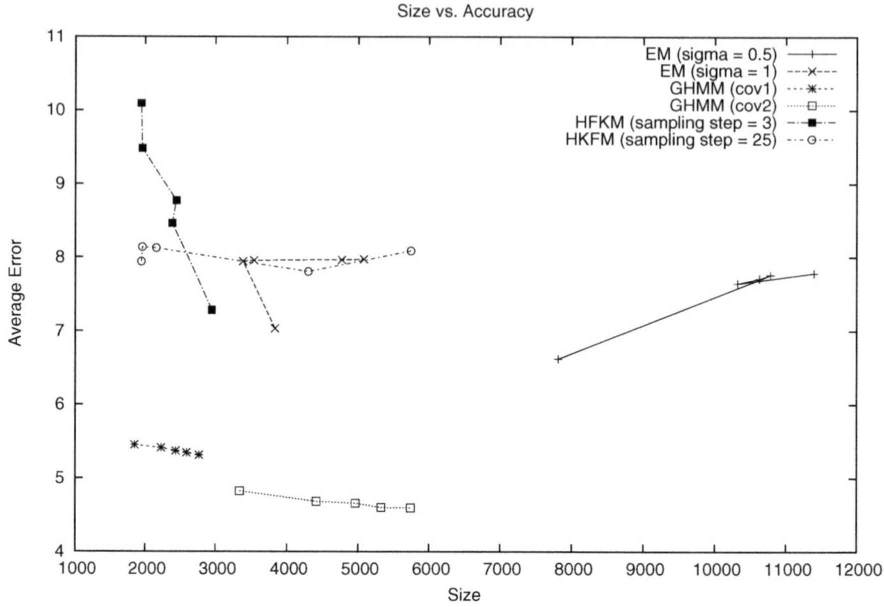

Fig. 8.16 Synthetic data: model size vs. prediction accuracy.

8.6 Discussion

In this chapter, we presented the results of applying our lifelong learning approach to the data sets introduced in chapter 7, with the intention of studying four points: a) the performance of the learning algorithm; b) the accuracy of the obtained predictions; c) the real time applicability of the whole approach; and d) the generality of our approach and its applicability to data sets having higher dimensions, or containing cyclic motion.

We illustrated with examples how our algorithm behaves with both data sets, illustrating along the way some of the problems which arose in the implementation. We explained the need for a performance measure, and proposed the average prediction error to evaluate our algorithm. We have also presented the quantitative results we have obtained using different sets of parameters for every data set. We have compared our approach against two state-of-the-art techniques. In these experiments, our algorithm has performed consistently better than the other two approaches by providing more accurate predictions with smaller models. Even if further experiments would be necessary to reach definitive conclusions, we consider these results to be very encouraging. Moreover, our experiments also show that our approach is able to perform lifelong learning and prediction in real-time even for large datasets.

We would like to mention that we have not reported in this chapter the preliminary results we have obtained with the second real data set described in Chapter 7, the interested reader can find them in Vasquez [2007].

It is worth noting that the code of our algorithm has been made available as a C^{++} library, together with some datasets. It can be found at: `http://www.dizanvasquez.net/ghmm`.

Part V
Conclusion

Chapter 9
Conclusions and Future Work

9.1 Conclusions

In this book we focused on the problem of predicting the motion of mobile objects which are able to modify their trajectories at will, as is the case for persons, vehicles and the like. Our work builds upon a family of approaches which are based on the idea that, for a given environment, objects follow typical motion patterns which may be observed consistently.

As discussed in ch. 3, most of these approaches operate in a *learn then predict* fashion [*eg* Walter et al., 1999, Makris and Ellis, 2002]: first, they learn motion patterns from sensor data, then, after learning has been performed, they predict further motion on the basis of learned patterns. As we have shown, this has a number of drawbacks, the most important being the inability to improve acquired knowledge when a previously unknown motion pattern is observed.

In contrast, we have aimed to find an lifelong learning capable approach which continuously refines its knowledge using the same observations that constitute the input for prediction. We have decided to base our work on Hidden Markov Model, a probabilistic framework, which is very popular in the literature [*eg* Koller-Meier and Van Gool, 2001, Bennewitz et al., 2005]. However, unlike these approaches – which rely on clustering algorithms to identify motion patterns – we have aimed to integrate the whole learning process into an HMM parameter and structure learning algorithm.

Our approach if based on the idea that a single HMM may represent multiple motion patterns by defining an appropriate graph structure. This idea is already present in other approaches, where different motion patterns correspond to different non-connected components of the HMM structure. Therefore, learning motion patterns may be seen as an HMM structure learning problem, with the additional constraint of working incrementally to allow for lifelong learning.

Unfortunately, as we have shown in ch. 4, no structure learning algorithm in the literature seems to be fit to our problem. This has led us to develop a novel incremental HMM structure and parameter learning algorithm.

A.D. Vasquez: Incre. Learn. for Motion Predict. of Pedest. & Vehicles, STAR 64, pp. 131–135.
springerlink.com

The main difficulty behind incremental structure learning on HMMs comes from the huge size of the space of possible structures, which renders search-based approaches unpractical for our purposes. Hence, we have taken a different approach, since, in our case, an HMM is just a discrete approximation of a continuous phenomenon, it is possible to impose constraints to the HMM structure according to the topological properties of the observation space.

Given that HMM states correspond to discrete regions on the state space, it follows that continuous motion may only progress by traversing neighboring regions, which significantly limits the number of possible transitions between states. This observation has suggested the use of a topology learning network known as the Instantaneous Topological Map to learn the HMMs topology. For this network, learning is both incremental and adaptive, hence, our structure learning algorithm also has this properties.

Of course, a complete HMM learning algorithm should also estimate the model's parameters, thus, we have interleaved our topology learning algorithm with an incremental HMM parameter learning algorithm based on the ideas developed by Neal and Hinton [1998]. This results in a general procedure for learning the structure and parameters of an HMM, which is presented in ch. 5. In contrast with traditional HMMs, the number of discrete states on the model changes with time – mostly growing – thus the name "Growing Hidden Markov Models".

As we show, GHMMs, are well suited for the common case in which the model is used as an approximate inference tool for continuous state spaces. However, in order to apply them successfully to the motion prediction problem, further assumptions should be made. We have explained the problem in 6 and proposed a solution: integrating the final state of the object into the state vector. Since this information is available during learning this may be regarded as a semi-supervised learning approach, in which the final destination or goal constitutes a class label.

We have implemented our approach and demonstrated with experiments on real and synthetic data (chapters 7 and 8) that our approach:

- Is able to learn motion patterns and represent them by incrementally adapting the structure of the GHMM.
- Is able to perform state estimation, as well as prediction for an arbitrary time horizon, using a probabilistic belief representation.
- Is able to perform early identification of an object's destination.
- It is able to perform both learning and prediction in real time even on complex environments and situations.
- Is defined in terms of intuitive parameters which do not require prior knowledge on the form of the environment or the number of motion patterns to be learned.
- Provides better or equivalent results to other state-of the art approaches while yielding smaller models.

We consider that these results are encouraging, however, we are aware that pattern based motion prediction techniques are still at a very early state. In consequence, during our experiments and – in general – our research, we have found several areas for improvement, which will be discussed on the next section.

9.2 Future Work and Possible Extensions

We see two general directions to extend the work presented in this book. The first one consists in extending the approach towards a higher level of abstraction, specifically by taking into account the possible interactions between objects. The second one consists in improving the low-level, making the approach more robust to the limitations of current video trackers and sensors in general.

9.2.1 High-Level Extensions

Currently, our approach does not take into account the possibility that an objects modifies its trajectory in response to the motion of another object. Here, we will discuss how our approach may be extended to handle some specific situations of this type.

A situation which is relatively simple to represent in our approach is the case where there are semi-static obstacles in the environment, *ie* objects which are not able to move freely, but may be in one of two possible states (*eg* 'block' or 'not block') which condition the choice of paths that other object may follow. This is the case, for example, of doors, which may be either 'open' or 'closed', or parking places, which may be either 'free' or 'occupied'.

Assuming that the information on the state of these obstacles is available via a sensor, extending inference is relatively simple. The key idea is that, when an object is is in a 'block' state, it hinders other object's access to certain states. We will illustrate this idea on the T-shaped environment presented in fig. 9.1(a).

A simple HMM structure for this environment is presented in fig. 9.1(b). When the door is closed, it "occupies" the grayed-out state. Although in our example there is only one such state, in general we will use the notation $occ(obstacle)$ to denote the set of states that a static object occupies when it is in its block state.

In order to perform inference, we augment our model with a variable $S_{door} \in \{open, closed\}$, indicating the state of the door, modifying the JPD as follows:

$$P(S_t\ S_{t-1}\ O_t\ S_{door}) = P(S_{t-1}\ S_{door})P(S_t \mid S_{t-1}\ S_{door})P(O_t \mid S_t\ S_{door}) \tag{9.1}$$

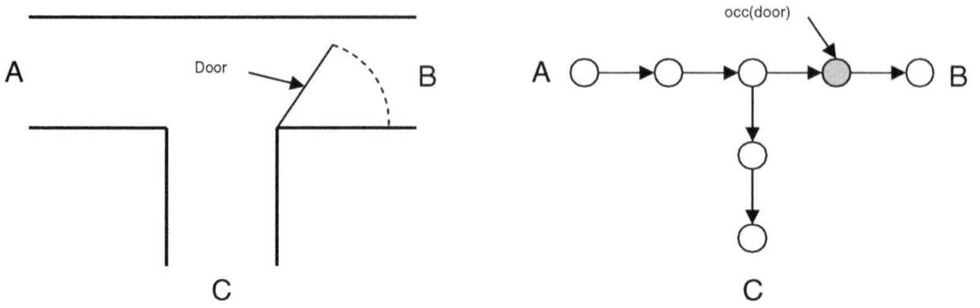

(a) Environment (b) HMM structure

Fig. 9.1 Example of an environment with a semi-static obstacle (door). The structure assumes that motion starts from point "A".

Since, when the door is closed, no object can be at the *occ*(*door*), it follows that:

$$P(S_{t-1}\ S_{door}) = \begin{cases} 0 & \text{if } S_{t-1} \in occ(door), S_{door} = closed \\ P(S_{t-1}) & \text{otherwise} \end{cases} \tag{9.2}$$

For analogous reasons, transitions to an occupied state are forbidden:

$$P(S_t \mid S_{t-1}\ S_{door}) = \begin{cases} 0 & \text{if } S_t \in occ(door), S_{door} = closed \\ P(S_t \mid S_{t-1}) & \text{otherwise} \end{cases} \tag{9.3}$$

and there is a null probability that an observation comes from an occupied state:

$$P(O_t \mid S_t\ S_{door}) = \begin{cases} 0 & \text{if } S_t \in occ(door), S_{door} = closed \\ P(O_t \mid S_t) & \text{otherwise} \end{cases} \tag{9.4}$$

These definitions may be easily extended for multiple semi-static objects, and used to perform inference using standard methods. Information about semi-static objects may be supplied by an expert, but fully unsupervised learning becomes considerably harder: not only it is necessary to identify semi-static objects, but also the states their block.

A much more challenging extension, would be taking into account interactions between fully dynamic objects. Normally, this would imply modeling the joint state of all the objects that are present in the environment at the same time, but this is clearly intractable. Instead, we believe that it would be interesting to investigate how to integrate factored solutions, like CHMM [Brand et al., 1997] or DML-HMM [Gong and Xiang, 2003, Xiang and Gong, 2006].

Another alternative would be to integrate our approach with behavioral models like the one proposed by Oliver et al. [2000], which are able to infer high level concepts like "object x is approaching object y", but are unable to determine metric information which is useful for motion planning. It would be very interesting to adapt our structure learning to provide a 'metric layer' for this kind of approaches.

9.2.2 Low-Level Extensions

As most pattern based approaches, our approach is based on the assumption that input is available in the form of complete observation sequences for every object. Nevertheless, as discussed in chapter 7, this assumption does not hold in practice, at least when using visual trackers. As a consequence, many motion patterns which do not always correspond to the actual object behavior are learned, unnecessarily increasing the model's size and significantly affecting prediction accuracy. This problem has been addressed in the literature by implementing "anomaly detection" algorithms [*eg* Hu et al., 2006] to filter out abnormal trajectories, unfortunately, they rely on having complete data sets and are designed to work in batch mode.

We believe that it would be better to explore ways to increase the robustness of pattern based approaches by relaxing this "complete trajectory" assumption. We think that, in order to achieve this, it will be necessary to get closer to the tracking system. A possibility is to study means to integrate our approach into the prediction step of the tracker.

Indeed, some steps have been already taken in this direction. Two noteworthy examples are: [Liao et al., 2003] and Bennewitz et al. [2005]. However, there is still a lot of work to do: the first approach assumes that a map of the environment is known *a priori*, while the second one assumes that the learning if performed on "complete trajectories" which is a circular problem, unless a more reliable sensor (*eg* a laser scanner) is used for learning.

Appendix A
Vector Quantization and Topology Representing Networks

Vector quantization is a data compression technique originally developed in the context of signal processing but widely used in very diverse technical domains. The idea of vector quantization is to encode a continuous D-dimensional input data manifold M by employing a finite set $C = \{c_1, \cdots, c_K\}$ of reference D-dimensional vectors. A point x of the manifold is represented using the element of C which is closest to it according to a given distance measure $d(x, c_i)$, such as the square error, or the Euclidean distance.

This procedure induces an implicit partition of the manifold in a number of subregions

$$\mathcal{V}_j = \{x \in M \mid d(x - c_j) \leq d(x - c_i) \forall i\} \tag{A.1}$$

called Voronoi regions, such that every input vector that is inside a Voronoi region $\mathcal{V}_j$ is described by the corresponding reference vector c_j.

The goal of a vector quantization algorithm is to find values for the reference values in order to minimize the mean quantization error, also known as the *distortion*:

$$E = \sum_{i=1}^{K} \int_{x \in \mathcal{V}_i} d(x, c_i) P(x) dx \tag{A.2}$$

Since, in most cases, the form of the manifold is unknown, the error may not be computed directly, instead, it is estimated from a data set consisting of $|X|$ samples, or input vectors:

$$\hat{E} = \frac{1}{|X|} \sum_{i=1}^{K} \sum_{x_j \in \mathcal{V}_i} d(x_j, c_i) \tag{A.3}$$

which assumes a uniform prior over input vectors.

The most widely used vector quantization algorithm is the k-means or Lloyd algorithm [Lloyd, 1957, Linde et al., 1980], which may be regarded as a version of the Expectation-Maximization algorithm, where a hard ownership criterion is used (*ie* every data element is assigned to one cluster only).

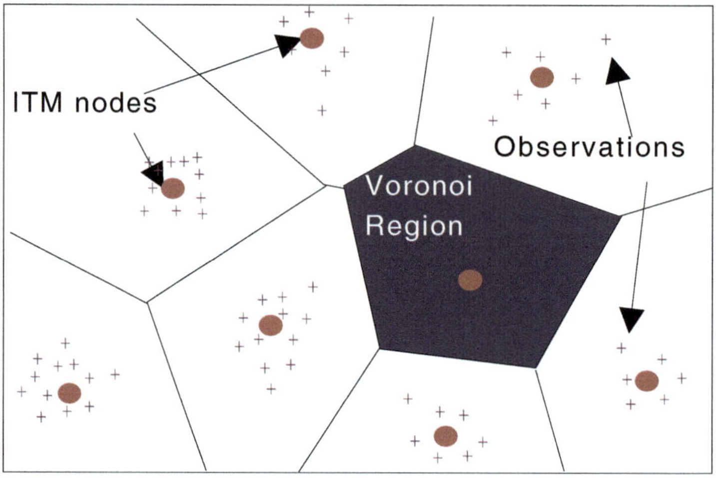

Fig. A.1 Example of a partition in Voronoi regions: there are some 2-dimensional input vectors (red crosses). Reference vectors are represented by big points and Voronoi regions are indicated by blue boundary lines, the set of all those boundaries is called a Voronoi graph.

Although standard k-means is guaranteed to converge to a local minimum of the distortion, it has a number of drawbacks: a) the number of reference vectors to be found should be known *a priori*; b) the quality of the output is highly dependent on initialization; c) the fact that it is a batch algorithm makes it too expensive for very big data sets.

Even if the last drawback has been addressed by MacQueen [1967], who proposed an incremental version of the algorithm – known as on-line k-means – the other two problems remain difficult and have motivated the research of alternative vector quantization approaches.

One of such approaches, which is particularly interesting, is the use of Kohonen Networks, or Self-organizing map (SOM) [Kohonen, 1995]. The main difference between the SOM and other classic vector quantization algorithms is that, in the SOM, there are links between reference vectors (or units, as they are known in the SOM jargon) so that they form a network. Links describe a neighborhood relationship between units and define a topology over the entire network. This topology is defined *a priori* to form a chain – for a one dimension Kohonen map – or a grid – for the two dimensional case.

In addition to vector quantization, the SOM learning algorithm ensures that, after learning, points that are close in the input data manifold M will be associated to the same unit, or to units which are close in the chain or grid. Hence, network links effectively encode additional information – called a topographic mapping – on the similarity of the represented input data. This additional information is useful in applications such as motion planning [Krose and Eecen, 1994], and speech processing [Kohonen, 1988].

Moreover, SOMs have been shown to reduce the initialization problem of k-means [Bacao et al., 2005].

However, Kohonen networks introduce a new problem: to obtain an optimal topographical mapping and minimize the quantization error, it is necessary that the topology of the network matches that of the represented manifold. In the case of Kohonen networks this requires prior knowledge about the topological structure of the manifold, which is often unavailable. This problem has motivated a new family of approaches known as Topology Representing Networks, which are able to learn the manifold's structure while performing vector quantization.

A.0.1 Topology Representing Networks

Topology Representing Networks are based on the idea of connecting neighbor Voronoi regions – *eg* regions with a common border – with edges, called delaunay's edges. The set of all edges constitutes the dual of the Voronoi graph and is known as the Delaunay's triangulation (fig. A.2).

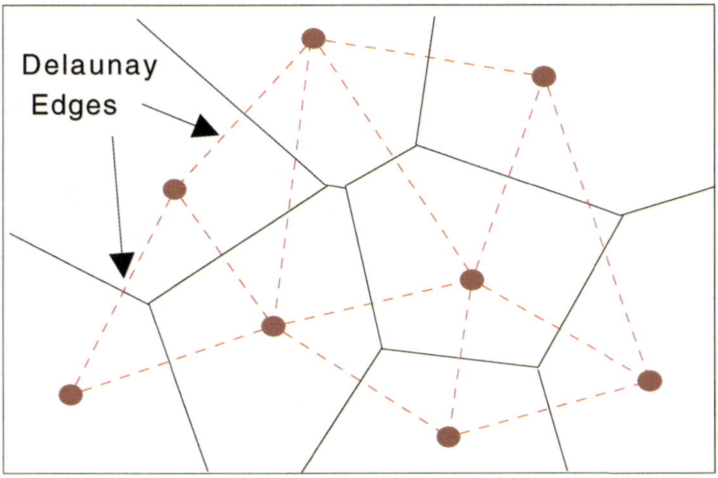

Fig. A.2 Voronoi graph (blue) and delaunay triangulation (green).

Topology-representing networks (TRN)s represent the network's topology with a subset of the Delaunay's triangulation. The edges are learned by using the competitive hebbian rule – also known as hebbian learning – proposed by Martinetz and Schulten [1991], which consists in creating a new edge between two units every time that, for a given input, they are the two closest units to that input and they are not already linked (fig. A.3).

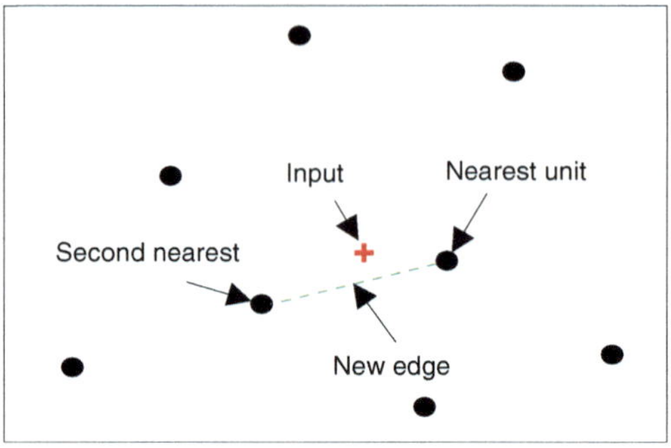

Fig. A.3 Hebbian learning

Although the first TRNs in the literature [*eg* Martinetz and Schulten, 1991]. worked only with a fixed number of units, later approaches, such as the Growing Neural Gas (GNG) [Fritzke, 1995], and the Grow When Required (GWR) Networks [Marsland et al., 2002], evolved as adaptive structures which are able to insert or even delete units during learning.

A common feature of these TRNs is that they adapt reference vectors by applying constant factors (*ie* learning rates) meaning that convergence to a minimum of distortion is not guaranteed, but this is not necessarily a drawback, since it allows the network to learn permanently. Indeed, this characteristic allows the network to show an adaptive behavior, since it is able to "forget" what it has learned in the past.

The advantages and drawbacks of adaptive TRNs, when compared to conventional k-means or EM vector quantization approaches may be summarized as follows:

A.0.1.1 Advantages

1. They use incremental learning algorithms, thus, they may be applied to very large data sets or to on-line learning from data streams.
2. They do not require any prior knowledge on the number of reference vectors.
3. Due to their growth mechanism, which uses observation data to initialize new state representations, they do not suffer the initialization problems of algorithms like k-means.
4. They are able to learn the data manifold's topology, even when it is mixed (*eg* it contains one, two and three dimensional submanifolds).
5. They are assumed to be a plausible model of how biological entities process information [cf Martinetz and Schulten, 1991].

A.0.1.2 Drawbacks

1. Not all of them converge to a minimum of the distortion (A.3) but, oscillate continuously around one.
2. Lacking or incomplete formal theories about their convergence and stability.

From our point of view, the drawbacks of TRNs are largely compensated by their advantages, which explains our decision of using a TRN – the Instantaneous Topological Map – as the basis for our learning algorithm.

Appendix B
Computing Fuzzy K-Means Cluster Representations

The original paper of Hu et al. [2006] presents the following expression to compute the jth cluster center $V_j(t)$:

$$V_j(t+1) = V_j(t) + \frac{\sum_{l=1}^{M} R_{lj}(t) \cdot (X_l - V_j(t))}{\sum_{l=1}^{M} R_{lj}(t)} \tag{B.1}$$

where M is the total number of trajectories, X_l is the lth trajectory, and $R_{lj}(t)$ is the estimated membership of trajectory l to cluster j. However, in our experiments using this expression has lead to bad clustering results. The problem seems to be that (B.1) contains a typo since it does not take into account the right *fuzzification coefficient* (see, for example [Kolen and Hutcheson, 2002]) whose value – judging from the other expressions in the paper – is two. The corrected equation has produced much better results, it is:

$$V_j(t+1) = V_j(t) + \frac{\sum_{l=1}^{M} R_{lj}^2(t) \cdot (X_l - V_j(t))}{\sum_{l=1}^{M} R_{lj}^2(t)} \tag{B.2}$$

Appendix C
Notation and Abbreviations

Here, we present a summary of the notation and abbreviations that we have used throughout this document. We have also adopted the standard convention of denoting random variables as capital letters, and values as lower-case letters. For logical propositions, we have used capital calligraphic characters, or expressions enclosed between brackets. When defining parametric forms, or working with linear algebra, vectors are represented as lower case letters to distinguish them from matrices, which are always capitalized.

List of Symbols

$C(\mathcal{A})$	The number (count) of cases in which the proposition $\mathcal{A}$ is true in experimental data.
D_V	The domain of V.
$\mathbf{G}(V;\ \mu,\ \Sigma)$	The gaussian distribution on V, with mean μ and covariance Σ.
O_t	The observation corresponding to time t.
S_t	The state of the object/system at time s.
t	The current discrete time step.
$\mathbf{U}_V$	The uniform probability distribution on V's domain.

List of Acronyms

AHMM .... Abstract Hidden Markov Models *page 37*
BIC .... Bayes Information Criterion *page 64*
CHMM .... Coupled Hidden Markov Models *page 37*
CPT .... Conditional Probability Table *page 19*
DML-HMM... Dynamically-multi-linked Hidden Markov Models *page 37*
EM .... Expectation-Maximization *page 20*

ESS .... Expected Sufficient Statistics *page 20*
GHMM .... Growing Hidden Markov Models *page 93*
GNG .... Growing Neural Gas *page 140*
GWR .... Grow When Required *page 140*
HMM .... Hidden Markov Models *page 34*
ITM .... Instantaneous Topological Map *page 72*
JPD .... Joint Probability Distribution *page 15*
MAP .... Maximum *a Posteriori* *page 63*
MDM .... Markov Dynamic Models *page 36*
ML .... Maximum Likelihood *page 63*
PDF .... Probability Density Function *page 15*
SOM .... Self-organizing map *page 138*
TRN .... Topology-representing networks *page 139*
VQ .... Vector Quantization *page 38*

References

Arulampalam, S., Maskell, S., Gordon, N., Clapp, T.: A tutorial on particle filters for on-line non-linear/non-gaussian bayesian tracking. IEEE Transactions on Signal Processing 50(2), 174–188 (2002)

Aurenhammer, F.: Voronoi diagrams - a survey of a fundamental geometric data structure. ACM Computing Surveys (CSUR) 23(3), 345–405 (1991)

Aurenhammer, F., Klein, R.: Voronoi diagrams, vol. V, pp. 201–290. Elsevier Science Publishing, Amsterdam (2000)

Bacao, F., Lobo, V., Painho, M.: Self-organizing maps as substitutes for k-means clustering. In: Proc. of the Int. Conf. on Computer Science, Atlanta (US), May 2005, pp. 476–483 (2005)

Baum, L., Petrie, T., Soules, G., Weiss, N.: A maximization technique occurring in the statistical analysis of probabilistic functions of markov chains. The Annals of Mathematical Statistics 41(1), 164–171 (1970)

Beckmann, N., Kriegel, H.-P., Schneider, R., Seeger, B.: The r*-tree: an efficient and robust access method for points and rectangles. In: Proc. of the 1990 ACM SIGMOD Int. Conf. on Management of data, Atlantic City (US), pp. 322–331. ACM Press, New York (1990)

Bellman, R.: Adaptive control processes: a guided tour. Princeton University Press, Princeton (1961)

Bennewitz, M., Burgard, W., Thrun, S.: Learning motion patterns of persons for mobile service robots. In: Proc. of the IEEE Int. Conf. On Robotics and Automation, Washington, USA, pp. 3601–3606 (2002)

Bennewitz, M., Burgard, W., Cielniak, G., Thrun, S.: Learning motion patterns of people for compliant robot motion. Internationl Journal of Robotics Research 24(1), 31–48 (2005)

Binsztok, H., Artières, T.: Learning model structure from data: an application to on-line handwriting. Electronic Letter on Computer Vision and Image Analyisis 5(2) (2005)

Boykov, Y., Kolmogorov, V.: An experimental comparison of min-cut/max-flow algorithms for energy minimization in vision. IEEE Trans. on Pattern Analysis and Machine Intelligence 26(9), 1124–1137 (2004)

Brand, M.: Structure learning in conditional probability models via an entropic prior and parameter extinction. Technical report, MERL a Mitsubishi Electric Research Laboratory (1998)

Brand, M., Kettnaker, V.: Discovery and segmentation of activities in video. IEEE Transactions on Pattern Analysis and Machine Intelligence 22(8), 844–851 (2000)

Brand, M., Oliver, N., Pentland, A.: Coupled hidden markov models for complex action recognition. In: Proc. of the 1997 Conf. on Computer Vision and Pattern Recognition, San Juan (PR), pp. 994–999 (1997)

Bruce, A., Gordon, G.: Better motion prediction for people-tracking. In: Proc. of the IEEE Int. Conf. on Robotics and Automation, New Orleans, US (April 2004)

Bui, H., Venkatesh, S., West, G.: Policy recognition in the abstract hidden markov models. Journal of Artificial Intelligence Research 17, 451–499 (2002)

Buzan, D., Sclaroff, S., Kollios, G.: Extraction and clustering of motion trajectories on video. In: Proc. of the Int. Conf. on Pattern Recognition, Cambridge, UK (2004)

Calinon, S., Billard, A.: Incremental learning of gestures by imitation in a humanoid robot. In: Proc. of the ACM/IEEE conf. on Human-Robot Interaction, pp. 255–262 (2007)

Caporossi, A., Hall, D., Reignier, P., Crowley, J.: Robust visual tracking from dynamic control of processing. In: International Workshop on Performance Evaluation of Tracking and Surveillance, Prague (CZ), May 2004, pp. 23–31 (2004)

Cassandra, A.R., Kaelbling, L.P., Kurien, J.A.: Acting under uncertainty: discrete bayesian models for mobile-robotnavigation, Osaka (JP), pp. 963–972 (1996)

Chang, C.C., Song, K.-T.: Environment prediction for a mobile robot in a dynamic environment. IEEE Transactions on Robotics and Automation 13(6), 862–872 (1997)

Cooper, G.F.: The computational complexity of probabilistic inference using bayesian belief networks. Artificial Intelligence 42(2-3), 393–405 (1990)

Cox, R.T.: Probability, frequency and reasonable expectation. American Journal of Physics (14), 1–13 (1946)

Dee, H., Hogg, D.: Detecting inexplicable behaviour. In: Proceedings of the British Machine Vision Conference, Kingston (UK), pp. 49–55 (2004)

Dee, H.-M.: Explaining Visible Behaviour. PhD thesis, University of Leeds (2005)

Dempster, N., Laird, A., Rubin, D.: Maximum likelihood from incomplete data via the EM algorithm. Journal of the Royal Statistical Society, Series B 9(1), 1–38 (1977)

Dennett, D.C.: The Intentional Stance. The MIT Press/Bradford Books, Cambridge (1987)

Dewdney, A.K., Vranch, J.K.: A convex partition of r 3 with applications to crum's problem and knuth's post-office problem. Utilitas Math. 12, 193–199 (1977)

Dixon, K.R., Dolan, J.M., Khosla, P.K.: Predictive robot programming: Theoretical and experimental analysis. Int. Journal of Robotics Research 23(9), 955–973 (2004)

Dwyer, R.A.: Higher-dimensional Voronoi diagrams in linear expected time. ACM Press, New York (1989)

Elnagar, A., Gupta, K.K.: Motion prediction of moving objects based on autoregressive model. IEEE Transaction on Systems, Man and Cybernetics, Part A 28(6), 803–810 (1998)

Filliat, D.: Cartographie et estimation globale de la position pour un robot mobile autonome. PhD thesis, Université Paris 6 (December 2001)

Foka, A.F., Trahanias, P.E.: Predictive autonomous robot navigation. In: Proc of the IEEE/RSJ Int. Conf. on Intelligent Robots and System, Lausanne (CH), October 2002, vol. 1, pp. 490–495 (2002)
Forney, G.D.: The viterbi algorithm. Proceedings of the IEEE 61(3), 268–278 (1973)
Freitag, D., McCallum, A.: Information extraction with hmm structures learned by stochastic optimization. In: Proc. of the Seventeenth Nat. Conf. on Artificial Intelligence and Twelfth Conf. on Innovative Applications of Artificial Intelligence, Austin, Texas, USA, July 2000, pp. 584–589. AAAI Press / The MIT Press (2000)
Friedman, N.: Learning belief networks in the presence of missing values and hidden variables. In: Proc. of the Fourteenth International Conference on Machine Learning, pp. 125–133 (1997)
Fritzke, B.: A growing neural gas network learns topologies. Advances in Neural Information Processing Systems (1995)
Gong, S., Xiang, T.: Recognition of group activities using dynamic probabilistic networks. In: Proceedings of the Ninth IEEE International Conference on Computer Vision, Washington (US), vol. 2, pp. 742–749 (2003)
Guttman, A.: R-trees: a dynamic index structure for spatial searching. In: Proc. of the 1984 ACM SIGMOD Int. Conf. on Management of data, Boston (US), pp. 47–57. ACM Press, New York (1984)
Guy, R.K.: Unsolved problems come of age. The American Mathematical Monthly 96(10), 903–909 (1989)
Han, K., Veloso, M.: Physical model based multi-objects tracking and prediction in robosoccer. In: Working notes of the AAAI 1997 Fall Symposium on Model-directed Autonomous Systems, Boston (US) (November 1997)
Heckerman, D.: A tutorial on learning with bayesian networks. Technical Report MSR-TR-95-06, Microsoft Research, Redmond, US (1995)
Hoogs, A., Perera, A.A.: Video activity recognition in the real world. In: Proc. of the Twenty-Third AAAI Conf. on Artificial Intelligence, Chicago, US, pp. 1551–1554 (2008)
Howard, R.A.: Dynamic Programming and Markov Process. MIT Press, Cambridge (1960)
Hu, W., Xie, D., Tan, T.: A hierarchical self-organizing approach for learning the patterns of motion trajectories. IEEE Transactions on Neural Networks 15(1), 135–144 (2004)
Hu, W., Xie, D., Tieniu, T., Maybank, S.: Learning activity patterns using fuzzy self-organizing neural network. IEEE Trans. on Systems, Man and Cybernetics 34(3), 1618–1626 (2004b)
Hu, W., Xiao, X., Fu, Z., Xie, D., Tan, T., Maybank, S.: A system for learning statistical motion patterns. IEEE Transactions on Pattern Analysis and Machine Intelligence 28(9), 1450–1464 (2006)
Igel, C., Suttorp, T., Hansen, N.: A computational efficient covariance matrix update and a (1+1)-cma for evolution strategies. In: Proc. of the 8th annual conf. on Genetic and Evolutionary Computation, pp. 453–460. ACM Press, New York (2006)
Irony, T., Singpurwalla, N.: Noninformative priors do not exist: A discussion with jose m. bernardo. Journal of Statistical Infererence and Planning 65, 159–189 (1997)

Rittscher, A.H.J., Blake, A., Stein, G.: Mathematical modelling of animate and intentional motion. Philosophical transactions-Royal Society of London. Biological sciences 358(1431), 475–490 (2003)

Jain, A., Murty, M., Flynn, P.: Data clustering: A review. ACM Computing Surveys 31, 265–322 (1999)

Jaynes, E.T.: Probability theory- the logic of science. Unpublished manuscript (1995), http://bayes.wustl.edu

Jockusch, J., Ritter, H.: An instantaneous topological map for correlated stimuli. In: Proc. of the International Joint Conference on Neural Networks, Washington (US), July 1999, vol. 1, pp. 529–534 (1999)

Johnson, N., Hogg, D.: Learning the distribution of object trajectories for event recognition. In: Proc. of the British Machine Vision Conference, September 1995, vol. 2, pp. 583–592 (1995)

Juang, B.-H., Levinson, S.E., Sondhi, M.M.: Maximum likelihood estimation for multivariate mixture observations of markov chains. IEEE Transactions on Information Theory 32(2), 307–309 (1986)

Julier, S., Uhlmann, J.: A new extension of the kalman filter to nonlinear systems. In: Int. Symp. Aerospace/Defense Sensing, Simul. and Controls, Orlando, US (1997)

Junejo, I., Javed, O., Shah, M.: Multi feature path modeling for video surveillance. In: Proc. of the 17th conference of the International Conference on Pattern Recognition (ICPR), pp. 716–719 (2004)

Kalman, R.: A new approach to linear filtering and prediction problems. Journal of Basic Engineering 82, 35–45 (1960)

Kaufman, L., Rousseeuw, P.J.: Finding Groups in Data: An Introduction to Cluster Analysis. Wiley Series In Probability And Mathematical Statistics. John Wiley and Sons, Inc., Chichester (1989)

King, B.: Step-wise clustering procedures. Journal of the American Statistical Association 69, 86–101 (1967)

Kohonen, T.: The 'neural' phonetic typewriter. Computer 21(3), 11–22, (1988), http://dx.doi.org/10.1109/2.28

Kohonen, T.: Self-Organizing Maps. Springer Series in Information Sciences, vol. 30. Springer, Heidelberg (1995); Second Extended Edition (1997)

Kolen, J.F., Hutcheson, T.: Reducing the complexity of the fuzzy c-means algorithm. IEEE Transactions on Fuzzy Systems 10(2), 263–267 (2002)

Koller-Meier, E.B., Van Gool, L.: Modeling and recognition of human actions using a stochastic approach. In: 2nd European Workshop on Advanced Video-Surveillance Systems, Kingston, UK (September 2001)

Krose, B.J.A., Eecen, M.: A self-organizing representation of sensor space for mobile robot navigation. In: Proc. of the IEEE/RSJ/GI Int. Conf. on Intelligent Robots and Systems, Munich (GR), vol. 1, pp. 9–14 (1994)

Kruse, E., Wahl, F.: Camera-based observation of obstacle motions to derive statistical data for mobile robot motion planning. In: Proc. of the IEEE Int. Conf. on Robotics and Automation, Leuven (BE), May 1998, pp. 662–667 (1998)

Kruse, E., Gutsche, R., Wahl, F.: Estimation of collision probabilities in dynamic environments for path planning with minimum collision probability. In: Proc. of the IEEE/RSJ Int. Conf. on Intelligent Robots and Systems, pp. 1288–1295 (1996)

Kruse, E., Gusche, R., Wahl, F.M.: Acquisition of statistical motion patterns in dynamic environments and their application to mobile robot motion planning. In: Proc. of the IEEE/RSJ Int. Conf. on Intelligent Robots and Systems, Grenoble, France, pp. 713–717 (1997)

Kuipers, B.: A hierarchy of qualitative representations for space. In: Freksa, C., Habel, C., Wender, K.F. (eds.) Spatial Cognition 1998. LNCS (LNAI), vol. 1404, pp. 337–350. Springer, Heidelberg (1998)

Kulic, D., Takano, W., Nakamura, Y.: Incremental learning, clustering and hierarchy formation of whole body motion patterns using adaptive hidden markov chains. Int. Journal of Robotics Research 27(7), 761–784 (2008)

Langley, P.: Order effects in incremental learning. In: Learning in Humans and Machines: Towards an Interdisciplinary Learning Science. Pergamon, Oxford (1995)

Lebeltel, O.: Programmation Bayéssiene des Robots. PhD thesis, Institut National Polytechnique de Grenoble (1999)

Levinson, S.E., Rabiner, L., Sondhi, M.M.: An introduction to the application of the theory of probabilistic functions of a markov process to automatic speech recognition. Bell System Technical Journal 62(4), 1035–1074 (1983)

Li, Y., Xu, L.-Q., Morphett, J., Jacobs, R.: An integrated algorithm of incremental and robust pca. In: Proc. of the Int. Conf. on Image Processing, vol. 1, pp. 245–248 (2003)

Liao, L., Fox, D., Hightower, J., Kautz, H., Schulz, D.: Voronoi tracking: Location estimation using sparse and noisy sensor data. In: Proc. of the IEEE/RSJ Int. Conf. on Intelligent Robots and Systems, Las Vegas, US (2003)

Linde, Y., Buzo, A., Gray, R.: An algorithm for vector quantizer design. IEEE Transactions on Communications COM-28, 84–95 (1980)

Liu, P.X., Meng, M., Hu, C.: On-line data-driven fuzzy clustering with applications to real-time robotic tracking. In: 2004 IEEE Int. Conf. on Robotics and Automation, New Orleans, USA, April 2004, pp. 5039–5044 (2004)

Lloyd, S.P.: Least squares quantization in pcm's. Bell Telephone Laboratories Paper (1957)

Lockwood, P., Blanchet, M.: An algorithm for the dynamic inference of hidden markov models (dihmm). In: Proc. IEEE Int. Conf. on Acoustics, Speech, and Signal Processing, Minneapolis, US, April 1993, vol. 2 (1993)

MacQueen, J.: Some methods for classification and analysis of multivariate observations. In: Cam, L.L., Neyman, J. (eds.) Proceedings of the 5th Berkeley Symposium on Mathematical Statistics and Probability, vol. 1, pp. 281–297. University of California Press (1967)

Magee, D.: Tracking multiple vehicles using foreground, background and shape models. Image and Vision Computing 22, 143–155 (2004)

Makris, D., Ellis, T.: Finding paths in video sequences. In: Proc. of the British Machine Vision Conference, pp. 263–272 (2001)

Makris, D., Ellis, T.: Spatial and probabilistic modelling of pedestrian behavior. In: Proc. of the British Machine Vision Conference, Cardiff, UK, pp. 557–566 (2002)

Mann, T.: Numerically stable hidden markov model implementation. An HMM scaling tutorial (2006)

Manning, C.D., Schutze, H.: Foundations of Statistical Natural Language Processing. The MIT Press, Cambridge (1999)

Marsland, S., Shapiro, J., Nehmzow, U.: A self-organizing network that grows when required. Neural Networks (2002)

Martinetz, T., Schulten, K.: A "neural-gas" network learns topologies. Artificial Neural Networks I, 397–402 (1991)

Meila, M., Jordan, M.I., Kaelbling, L.P.: Learning with mixtures of trees. Journal of Machine Learning Research 1(1), 1–48 (2001)

Minka, T.P.: From hidden markov models to linear dynamical systems. Technical report, MIT (1999)

Minnen, D.C., Wren, C.R.: Finding temporal patterns by data decomposition. In: Proc. of the Sixth IEEE Int. Conf. on Automatic Face and Gesture Recognition, pp. 608–613 (2004)

Murphy, K.P.: Dynamic Bayesian Networks: Representation, Inference and Learning. PhD thesis, University of California, Berkeley, USA (2002)

Neal, R.M., Hinton, G.E.: A new view of the em algorithm that justifies incremental, sparse and other variants. In: Jordan, M.I. (ed.) Learning in Graphical Models, pp. 355–368. Kluwer Academic Publishers, Dordrecht (1998)

Oliver, N.M., Rosario, B., Pentland, A.P.: A bayesian computer vision system for modeling human interactions. IEEE Transactions on Pattern Analysis and Machine Intelligence 22(8), 831–843 (2000)

Osentoski, S., Manfredi, V., Mahadevan, S.: Learning hierarchical models of activity. In: Proc. of the IEEE/RSJ Int. Conf. on Intelligent Robots and Systems, Sendai, Japan (2004)

Pearl, J.: Probabilistic Reasoning in Intelligent Systems: Networks of Plausible Inference. Morgan Kauffman, San Francisco (1988)

Pentland, A., Liu, A.: Modeling and prediction of human behavior. Neural Computation 11(1), 229–242 (1999)

Rabiner, L.R.: A tutorial on hidden markov models and selected applications in speech recognition. In: Readings in speech recognition, pp. 267–296 (1990)

Rao, R.P.N.: Robust kalman filters for prediction, recognition, and learning. Technical Report TR645, The University of Rochester (1996), `http://citeseer.ist.psu.edu/rao96robust.html`

Reif, J., Sharir, M.: Motion planning in the presence of moving obstacles. In: Symp. on the Foundations of Computer Science, Portland, US, October 1985, pp. 144–154 (1985)

Madhavan, R., Schlenoff, C.: Moving object prediction for off-road autonomous navigation. In: Proceedings of the SPIE Aerosense Conference, Orlando, FL, USA (2003)

Rosales, R., Sclaroff, S.: Improved tracking of multiple humans with trajectory prediction and occlusion modeling. In: IEEE CVPR Workshop on the Interpretation of Visual Motion (1998)

Schwarz, G.: Estimating the dimension of a model. The Annals of Statistics 6(2), 461–464 (1978)

Seymore, K., McCallum, A., Rosenfeld, R.: Learning hidden markov model structure for information extraction. In: AAAI 99 Workshop on Machine Learning for Information Extraction, Orlando, US (1999)

Singer, Y., Warmuth, M.K.: Training algorithms for hidden markov models using entropy based distance functions. In: Advances in Neural Information Processing Systems 9, NIPS, Denver, CO (USA), December 2-5, pp. 641–647. MIT Press, Cambridge (1996)

Stauffer, C., Grimson, E.: Learning patterns of activity using real-time tracking. IEEE Transactions on Pattern Analysis and Machine Intelligence 22(8), 747–757 (2000)

Stolcke, A., Omohundro, S.: Hidden markov model induction by bayesian model merging. In: Hanson, S.J., Cowan, J.D., Giles, C.L. (eds.) Advances in Neural Information Processing Systems, Denver, USA, vol. 5, pp. 11–18. Morgan Kaufmann, San Mateo (1993)

Stolcke, A., Omohundro, S.M.: Best-first model merging fr hidden markov model induction. Technical Report TR-94-003, International Computer Science Institute, Berkeley, US, 1994 (1994)

Stone, L., Barlow, C., Corwin, T.: Bayesian Multiple Target Tracking. Artech House (1999)

Sumpter, N., Bulpitt, A.: Learning spatio-temporal patterns for predicting object behaviour. Image and Vision Computing 18(9), 697–704 (2000)

Tadokoro, S., Hayashi, M., Manabe, Y., Nakami, Y., Takamori, T.: Motion planner of mobile robots which avoid moving human obstacles on the basis of stochastic prediction. In: IEEE International Conference on Systems, Man and Cybernetics, pp. 3286–3291 (1995)

Thrun, S., Burgard, W., Fox, D.: Probabilistic Robotics. MIT Press, Cambridge (2005)

Vasko, R.C., El-Jaroudi, A., Boston, J., Rudy, T.E.: Hidden markov model topology estimation to characterize the dynamic structure of repetitive lifting data. In: Proc. of the 19th Annual Int. Conf. of the IEEE Engineering in Medicine and Biology Society, Chicago, US, October 1997, pp. 1725–1728 (1997)

Vasquez, D.: Incremental Learning for Motion Prediction of Pedestrians and Vehicles. PhD thesis, Institut National Polytechnique de Grenoble, Grenoble, FR (February 2007), http://emotion.inrialpes.fr/bibemotion/2007/Vas07

Vasquez, D., Fraichard, T.: Motion prediction for moving objects: a statistical approach. In: Proc. of the IEEE Int. Conf. on Robotics and Automation, New Orleans, LA (US), pp. 3931–3936, (April 2004), http://emotion.inrialpes.fr/bibemotion/2004/VF04

Viterbi, A.J.: Error bounds for convolutional codes and an asymptotically optimum decoding algorithm. IEEE Transactions on Information Theory IT-13(2), 260–269 (1967)

Walter, M., Psarrow, A., Gong, S.: Learning prior and observation augmented density models for behaviour recognition. In: Proc. of the British Machine Vision Conference, pp. 23–32 (1999)

Wang, X., Tieu, K., Grimson, E.: Learning semantic models by trajectory analysis. Technical report, Massachussets Institute of Technology (2006)

Xiang, T., Gong, S.: Beyond tracking: Modelling activity and understanding behaviour. International Journal of Computer Vision (2006)

Xie, X.L., Beni, G.: A validity measure for fuzzy clustering. IEEE Transactions on Pattern Analysis and Macine Intelligence 13(8), 841–847 (1991)

Yu, H., Su, T.: Destination driven motion planning via obstacle motion prediction and multi-state path repair. Journal of Intelligent and Robotic Systems 36(2), 149–173 (2003)

Zhu, Q.: A stochastic algorithm for obstacle motion prediction in visual guidance of robot motion. In: The IEEE International Conference on Systems Engineering, pp. 216–219 (1990)

Springer Tracts in Advanced Robotics

Edited by B. Siciliano, O. Khatib and F. Groen

Further volumes of this series can be found on our homepage: springer.com

Vol. 63: Vanderborght, B.;
Dynamic Stabilisation of the
Biped Lucy Powered by Actuators
with Controllable Stiffness
281 p. 2010 [978-3-642-13416-6]

Vol. 62: Howard, A.; Iagnemma, K.;
Kelly, A. (Eds.):
Field and Service Robotics
511 p. 2010 [978-3-642-13407-4]

Vol. 61: Mozos, Ó.M.
Semantic Labeling of Places with Mobile Robots
134 p. 2010 [978-3-642-11209-6]

Vol. 60: Zhu, W.-H.
Virtual Decomposition Control – Toward Hyper
Degrees of Freedom Robots
443 p. 2010 [978-3-642-10723-8]

Vol. 59: Otake, M.
Electroactive Polymer Gel Robots – Modelling
and Control of Artificial Muscles
238 p. 2010 [978-3-540-23955-0]

Vol. 58: Kröger, T.
On-Line Trajectory Generation in Robotic
Systems – Basic Concepts for Instantaneous
Reactions to Unforeseen (Sensor) Events
230 p. 2010 [978-3-642-05174-6]

Vol. 57: Chirikjian, G.S.; Choset, H.;
Morales, M., Murphey, T. (Eds.)
Algorithmic Foundations
of Robotics VIII – Selected Contributions
of the Eighth International Workshop on the
Algorithmic Foundations of Robotics
680 p. 2010 [978-3-642-00311-0]

Vol. 56: Buehler, M.; Iagnemma, K.;
Singh S. (Eds.)
The DARPA Urban Challenge – Autonomous
Vehicles in City Traffic
625 p. 2009 [978-3-642-03990-4]

Vol. 55: Stachniss, C.
Robotic Mapping and Exploration
196 p. 2009 [978-3-642-01096-5]

Vol. 54: Khatib, O.; Kumar, V.;
Pappas, G.J. (Eds.)
Experimental Robotics:
The Eleventh International Symposium
579 p. 2009 [978-3-642-00195-6]

Vol. 53: Duindam, V.; Stramigioli, S.
Modeling and Control for Efficient Bipedal
Walking Robots
211 p. 2009 [978-3-540-89917-4]

Vol. 52: Nüchter, A.
3D Robotic Mapping
201 p. 2009 [978-3-540-89883-2]

Vol. 51: Song, D.
Sharing a Vision
186 p. 2009 [978-3-540-88064-6]

Vol. 50: Alterovitz, R.; Goldberg, K.
Motion Planning in Medicine: Optimization
and Simulation Algorithms for
Image-Guided Procedures
153 p. 2008 [978-3-540-69257-7]

Vol. 49: Ott, C.
Cartesian Impedance Control of Redundant
and Flexible-Joint Robots
190 p. 2008 [978-3-540-69253-9]

Vol. 48: Wolter, D.
Spatial Representation and
Reasoning for Robot
Mapping
185 p. 2008 [978-3-540-69011-5]

Vol. 47: Akella, S.; Amato, N.;
Huang, W.; Mishra, B.; (Eds.)
Algorithmic Foundation of Robotics VII
524 p. 2008 [978-3-540-68404-6]

Vol. 46: Bessière, P.; Laugier, C.;
Siegwart R. (Eds.)
Probabilistic Reasoning and Decision
Making in Sensory-Motor Systems
375 p. 2008 [978-3-540-79006-8]

Vol. 45: Bicchi, A.; Buss, M.;
Ernst, M.O.; Peer A. (Eds.)
The Sense of Touch and Its Rendering
281 p. 2008 [978-3-540-79034-1]

Vol. 44: Bruyninckx, H.; Přeučil, L.; Kulich, M. (Eds.)
European Robotics Symposium 2008
356 p. 2008 [978-3-540-78315-2]

Vol. 43: Lamon, P.
3D-Position Tracking and Control for All-Terrain Robots
105 p. 2008 [978-3-540-78286-5]

Vol. 42: Laugier, C.; Siegwart, R. (Eds.)
Field and Service Robotics
597 p. 2008 [978-3-540-75403-9]

Vol. 41: Milford, M.J.
Robot Navigation from Nature
194 p. 2008 [978-3-540-77519-5]

Vol. 40: Birglen, L.; Laliberté, T.; Gosselin, C.
Underactuated Robotic Hands
241 p. 2008 [978-3-540-77458-7]

Vol. 39: Khatib, O.; Kumar, V.; Rus, D. (Eds.)
Experimental Robotics
563 p. 2008 [978-3-540-77456-3]

Vol. 38: Jefferies, M.E.; Yeap, W.-K. (Eds.)
Robotics and Cognitive Approaches to Spatial Mapping
328 p. 2008 [978-3-540-75386-5]

Vol. 37: Ollero, A.; Maza, I. (Eds.)
Multiple Heterogeneous Unmanned Aerial Vehicles
233 p. 2007 [978-3-540-73957-9]

Vol. 36: Buehler, M.; Iagnemma, K.; Singh, S. (Eds.)
The 2005 DARPA Grand Challenge – The Great Robot Race
520 p. 2007 [978-3-540-73428-4]

Vol. 35: Laugier, C.; Chatila, R. (Eds.)
Autonomous Navigation in Dynamic Environments
169 p. 2007 [978-3-540-73421-5]

Vol. 34: Wisse, M.; van der Linde, R.Q.
Delft Pneumatic Bipeds
136 p. 2007 [978-3-540-72807-8]

Vol. 33: Kong, X.; Gosselin, C.
Type Synthesis of Parallel Mechanisms
272 p. 2007 [978-3-540-71989-2]

Vol. 30: Brugali, D. (Ed.)
Software Engineering for Experimental Robotics
490 p. 2007 [978-3-540-68949-2]

Vol. 29: Secchi, C.; Stramigioli, S.; Fantuzzi, C.
Control of Interactive Robotic Interfaces – A Port-Hamiltonian Approach
225 p. 2007 [978-3-540-49712-7]

Vol. 28: Thrun, S.; Brooks, R.; Durrant-Whyte, H. (Eds.)
Robotics Research – Results of the 12th International Symposium ISRR
602 p. 2007 [978-3-540-48110-2]

Vol. 27: Montemerlo, M.; Thrun, S.
FastSLAM – A Scalable Method for the Simultaneous Localization and Mapping Problem in Robotics
120 p. 2007 [978-3-540-46399-3]

Vol. 26: Taylor, G.; Kleeman, L.
Visual Perception and Robotic Manipulation – 3D Object Recognition, Tracking and Hand-Eye Coordination
218 p. 2007 [978-3-540-33454-5]

Vol. 25: Corke, P.; Sukkarieh, S. (Eds.)
Field and Service Robotics – Results of the 5th International Conference
580 p. 2006 [978-3-540-33452-1]

Vol. 24: Yuta, S.; Asama, H.; Thrun, S.; Prassler, E.; Tsubouchi, T. (Eds.)
Field and Service Robotics – Recent Advances in Research and Applications
550 p. 2006 [978-3-540-32801-8]

Vol. 23: Andrade-Cetto, J,; Sanfeliu, A.
Environment Learning for Indoor Mobile Robots – A Stochastic State Estimation Approach to Simultaneous Localization and Map Building
130 p. 2006 [978-3-540-32795-0]

Vol. 22: Christensen, H.I. (Ed.)
European Robotics Symposium 2006
209 p. 2006 [978-3-540-32688-5]

Vol. 21: Ang Jr., H.; Khatib, O. (Eds.)
Experimental Robotics IX – The 9th International Symposium on Experimental Robotics
618 p. 2006 [978-3-540-28816-9]

Vol. 20: Xu, Y.; Ou, Y.
Control of Single Wheel Robots
188 p. 2005 [978-3-540-28184-9]

Vol. 19: Lefebvre, T.; Bruyninckx, H.; De Schutter, J. Nonlinear Kalman Filtering for Force-Controlled Robot Tasks
280 p. 2005 [978-3-540-28023-1]

GPSR Compliance

The European Union's (EU) General Product Safety Regulation (GPSR) is a set of rules that requires consumer products to be safe and our obligations to ensure this.

If you have any concerns about our products, you can contact us on ProductSafety@springernature.com

In case Publisher is established outside the EU, the EU authorized representative is:

Springer Nature Customer Service Center GmbH
Europaplatz 3
69115 Heidelberg, Germany

Batch number: 10614853

Printed by Printforce, the Netherlands